ANECDOTES

A Tiff's
Life and Beyond

BY

TED BASS

Published by

MELROSE BOOKS

An Imprint of Melrose Press Limited
St Thomas Place, Ely
Cambridgeshire
CB7 4GG, UK
www.melrosebooks.co.uk

FIRST EDITION

Cover designed by Jeremy Kay

ISBN 978-1-908645-03-6

Printed and bound in Great Britain by:
Mimeo Ltd, Huntingdon, Cambridgeshire

FSC
www.fsc.org
MIX
From responsible
sources
FSC® C019549

CONTENTS

GLOSSARY OF TERMS

AFD	Admiralty Floating Dock
AWOL	Absent Without Leave
BMH	Base Medical Hospital
Boats	Submarines
Bootneck	Royal Marine
Chit	Piece of paper for permission/instruction
Chippy	Shipwright Artificer
Clacker	Pastry
Comforts	Special issue food for boats
CPO	Chief Petty Officer
Crushers	Leading Patrol Men – RN Policemen
Dabtoes	Denigratory term for seaman
Deckhead	Ceiling
Desiccate	Dry out using compressed air
DF	Duty free
DO	Divisional Officer-Head of Branch
DQs	Detention Quarters
EO	Engineer Officer
ERA	Engine Room Artificer
Fifth Five	To sign on for 5 more years after doing 22
Fizzer	On report – in trouble
Flat	Open area below deck
GBH	Grievous Bodily Harm
Greenies	Any member of the Electrical Department
Guvvy	Government work – illegal for self
Horoscope	The reading of – getting told off
Heads	Toilets
Jack	General term for sailor
Jack Dusty	Stores Assistant
Jimmy	First Lieutenant
Kegs	Trousers/underpants

Killick	Leading Hand – Name of Badge – Killick Anchor
Lagging	Kit/clothing
LCT	Landing Craft Tank
MAA	Master At Arms – Joss Man – God
MFV	Motor Fishing Vessel
MO	Medical Officer
Mob	Crowd/surround
Neaters	Undiluted rum – grog
Netting	Mesh area to store hammocks
Nettles	Multiple light lines at hammock ends
On	Owed favours called in
OD	Ordinary Seaman
Oggin	Sea-Ogwash
OOW	Officer Of The Watch
Oppo	Opposite watchkeeping number, friend, pal
Outside Wrecker	See Wrecker
Panel	Boat's diving, surfacing and mast control
Part Three	Third part of training/learning the boat
Piped	Bosun's pipe call – followed by instruction
Pompey	Portsmouth
Pussers	Royal Naval
RA	Rationed Ashore
Rattle	On report, in trouble
RBG	Rich Brown Gravy
REM	Radio/Radar Electrical Mechanician
Round the Buoy	Going for a second helping
RPO	Regulating Petty Officer (RN Police)
Scran Bag	Sculling gear collection
Scratcher	Bosun's Mate/Second Coxswain in boats
Second Dicky	Second in Command
Shitehawk	Seagull
Shufti	Take a look
Slop Chit	Piece of paper to allow issue of stores
Sprogging	Apprentice slaving for a Senior
Subby	Sub Lieutenant
TGs	Turbo Generators

Tiffs	Artificers – Tradesmen
Towny	Oppo – coming from the same area
Troop	To be put in the Rattle
Troops	Collective term for jack
US	Unserviceable
2-6	Heave, in cannon days nos 2 and 6 hauled out
UW Telephone	Under Water Telephone
Wrecker	Outside Tiff – sees to all but propulsion
Zeds	Sleep – knock out a few

HMS *FISGARD*

Scrumpy

For me it all started with an advertisement at the back of the *Radio Times* calling for apprentices in the Royal Navy, so I wrote off for details. What came back was great, but an examination had to be sat. I also had to have parental permission and Headmaster's permission to sit it. My Mother gave her permission and signed the consent form. The Headmaster was another kettle of fish, he positively sneered at me and as much as said I had no chance, which made me even more determined. To be quite honest I think his negative attitude was my inspiration to prove him wrong, but to this day I don't know if he took this course of action on purpose.

To cut a long story short I passed the exam along with Phil, another pupil from the same school but a year in front of me, who also passed. So we teamed up and went down to Portsmouth – HMS *Daedalus* – for the interview, practical exam and medical. I passed them all but poor old Phil failed the medical on a minor problem but was promised a place after treatment so it wasn't too bad, so we went to the canteen and were introduced to Scrumpy. We were told it was just cheap cider, and it tasted a bit rough, but it quenched the thirst. But nobody warned us about its potency and we both ended up wrecked, not a pretty sight. We were both dog rough in the morning, and were sent home to await instructions. Phil went off to have his problem fixed. I finally got my orders, and Mum was pleased for me and gave me a fag case and lighter, because she knew I smoked and off I went at the age of fifteen to sign on for twelve years from the age of eighteen, fifteen whole years, a lifetime when you're that age.

What a different world it was, a load of fifteen year olds thrown together from all walks of like to do an apprenticeship in the Royal Navy. The world was my oyster but I tended to stay away from the dreaded Scrumpy!!

Fisgard Boxing Team (Skive)

I joined the Royal Navy at age fifteen and never regretted it. I replied to a recruiting advert at the back of the *Radio Times* and got a reply inviting me to attend an interview down south at HMS *Daedalus*.

To cut a long story short I passed the interview and medical. I ended up on HMS *Fisgard* at Torpoint in Cornwall, the lowest of the low being bullied by the more senior hands, but I enjoyed the practical workshop programme and struggled through with the school work.

Once I had settled in and found out how to keep a low profile I began to relax and enjoy myself more. I also discovered that there were some good skives (schemes) to get involved with, but by far the best option was the ship's boxing team. You had to undergo very vigorous exercises, like long road runs, gym exercises, sparring and punch bag work, but the rewards were well worth it, such as being excused menial duties and issued with extra rations in the Mess Hall, and other useful perks.

Come the day of reckoning though, we were challenged to a match by HMS *Raleigh*'s Boxing Club, the Naval Establishment across the road from us, training sailors and stokers, who thought us Tiffs were worth taking down a peg or two as they knew we enjoyed accelerated promotion compared to them.

The stage was set for the tournament at *Raleigh*, so the audience were very anti-Tiff to start with. I was fighting at Mosquito weight as I was small and skinny and usually an opponent couldn't be found at my weight but this time they did. He was the skinniest thing I've ever seen and tall with it with arms down past his knees. To cut a long story short he knocked seven bells out of me because I couldn't even reach him let alone hit him.

The outcome of the tournament was that our team won by quite a margin, but needless to say, I resigned from the Boxing Club!

Fisgard **Booze (Punishment)**

As we Junior Apprentices progressed through training we encountered all sorts of temptations. I already smoked which my Mum knew as she gave me a full fag case and lighter as a parting gift.

Me and my mate Phil used to spend our school dinner money, him with the cash and me by flogging my free dinner ticket for slightly less than the going rate, at the local Selby Street chippy for a patty and chips each and then went next door to spend the remainder on five Woodbines. We then used to retire to the loft of a disused barn to consume our fry up and then get dizzy inhaling fag smoke.

I carried on smoking, but on an apprentice's pay, half of which went on laundry bills, very little was left for nutty and fags, so what I and a few others did was to split tailor-made fags, buy a packet of fag paper and re-roll the tobacco into twice as many skinnier fags.

But some of our more adventurous classmates turned to alcohol which was strictly illegal as we were all under age, and the penalty if caught was corporal punishment. This consisted of cuts, in other words caned on the arse through wet canvas kegs. I'm told that it hurts like hell and so tended to help the recipient desist and wait till the age of eighteen when it was legal to consume alcohol. He was also entitled, as we all were, to a tot each day, an eighth of a pint of finest dark rum.

Rum at tot time (noonish) in my day was almost a religion but could be abused and frequently was.

Fisgard **(Blood Poisoning)**

We arrived at *Fisgard*, all strangers, by RN Transport, having been collected at Guz (Plymouth) Railway Station, wearing our best civvy suits, and after a rather restless night in strange surroundings in a large dormitory with bedding supplied we were mustered on the parade ground and given a good talking to about the rules and regulations.

Then it's off to collect your Naval kit which includes a housewife (sewing kit) boots, work blues (shirts and trousers), socks, a one pint enamel mug and your name type (wooden one inch tall letters glued together to form your initial and surname), plus a few other bits and pieces, and finally your naval uniform suit, single breasted in rough serge material with black plastic buttons and a very stiff peaked cap. When you were dressed in this gear you looked and felt like a right prat, but you soon found out by copying the actions of the more senior classes and giving the hat at least a make-over by administering a good bashing to get it into a better shape.

If you wanted a decent going ashore or up homers suit you had to go to a Naval Tailor who was quite happy to make you a very smart double breasted doeskin rig, but at a price, and your Tailor was quite happy to allow you credit but at a price so you ended up making an allotment to him that kept you even more skint.

You also very quickly got to know the routine and learned what and what not to do or say, but the highlight of any day for me was the workshops. We started off with very basic hand tool work, hammer and chisels and an assortment of files complete with handles, a six inch steel rule and inside and outside callipers.

The first job we all tackled was a simple square mild steel block fitting into a square hole in a larger square cast iron block. It sounds easy, but there's a snag, you are issued with a length of round mild steel bar and a round cast iron disc with a small hole drilled through the middle of it, and a drawing of the finished job, complete with all dimensions.

We all ended up with very bloody left thumb knuckles or, right if left handed, through missing with the two and a half pound hammer.

Our instructor was a serving fleety Tiff (Fleet Air Arm Artificer) who was a real good hand and was a wiz at hand fitting and showed us a few tricks, like stuffing a wad of cotton waste against the vice securing bolt and soaking it in oil, where you dipped your chisel point into the wad to cut mild steel and see a pleasing puff of smoke as you peeled off nice curl of steel very easily and smoothly.

I was thoroughly enjoying myself in the workshop when disaster struck, caused by my No 8 trousers (dark blue work gear). Me and a friend were skylarking in the dormitory and in the melee I banged my right shin on a steel bed frame which took a lump out of my leg. My mate suggested I go to the sick bay but I thought better of it as we were fooling about and I thought we might both end up in the rattle.

Not having access to bandages or plasters and stupidly not wanting to invite attention I cleaned it up as best I could and carried on. Unbeknown to me was the fact that my new and as yet unwashed No 8 trousers had not very well fixed dark blue dye which somehow managed to contaminate my shin wound and over a few days I developed a quite nasty ulcerated leg and blood poisoning and ended up in sick bay for a week and got a right rollicking for not reporting sick earlier and I still have the scar to prove it!

Eventually on Discharge from Sick Bay and return to the workshop I found out that the time limit for the job I was half way through was fast approaching and to not finish would fail me, so I had to pull my finger out and rush to complete which I eventually did but I was not happy with the resultant fit and finish but the inspection placed me in the top quarter which wasn't bad, and I knew I could do better.

Stove Stoking

Imagine if you will, mid winter, very cold, thick snow, a twenty-plus bed dormitory and the only heating a pot bellied stove in the middle of the joint.

This stove was the focal point where Apprentices gathered to natter like old washer women and try to keep warm. This was the picture in a Senior Apprentices' dormitory. For us junior first term apprentices it was a matter of survival of the fittest almost.

We were at the Seniors' beck and call and one of our tasks was to keep their pot bellied stove fully stoked up. The snag was that the only fuel available was coke, and it was stored out in the open and consequently was soaking wet or frozen solid and there was only so much soggy coke you can get into the body of the stove and if the fire died down too much and tried to stoke it too well and too quickly the soggy coke would kill the fire completely.

One of the Seniors came up with the real solution. The stove must be fed slowly and gently so as not to overload it, but not from the front because there was no more room, but from up top.

Yes we had to go onto the roof and feed soggy coke a piece at a time down the chimney. A ladder was thieved from somewhere which got us onto the roof and then it was snow clearing time and a rope slung over the apex and secured on the other side. We then shinned up the rope with a bucket of coke to drop it piece by piece down the funnel so as not to overload the fire below. The first thing that happened when a fresh piece of coke was fed in, a plume of steam as it thawed and dried out then it fired up. If care was taken the whole of the flue could end up glowing red hot and almost doubled the heating capacity of the stove, while we almost froze to death on the outside and on the roof.

Welsh Rarebit

Again, as first term Artificer Apprentices, we were the lowest of the low and at everybody's beck and call. A sprigging system was in operation where a Senior Apprentice adopted one of us first termers as his slave. Some of the Seniors were fairly reasonable, while others were proper bastards.

I was one of the lucky ones who got picked by a reasonable and fair Senior, but I still had to break the rules for him by taking food out of the Mess Hall, which was strictly illegal and forbidden. Tastes of illegal snacks went in phases and fads by the seniors.

The latest favourite at this particular time was Welsh rarebit. In this case cheese on toast with brown sauce on top. It sounds easy but it wasn't at all, first of all the bread had to be smuggled out of the Mess Hall. You were officially allowed two slices of bread at tea time and if you wanted it for yourself, because you were always hungry, you would have to go round the buoy and hope the civilian staff didn't recognise you. If they did you would have to smuggle your own allowance out and donate it to your 'master' and go hungry yourself. The way you smuggled bread out was under your arm pit and make sure you'd showered and had on a clean shirt. You needed two slices, one as backup in case of 'accidents'. Then it was a case of 'borrowing' the cheese, enough for two portions, again just in case. The brown sauce was no problem, you just nicked the whole bottle!

Now you were nearly set and toasting the bread was no problem, a home- made toasting fork was employed to hold the slice up against the front grate of the pot bellied stove, the only source of heat in the dormitory.

Now comes the tricky bit, the cheese is sliced and covers the whole of the toast, if you left bare toast it would burn and possibly catch fire. The next stage was most important, the stove is stoked up and then riddled furiously to get rid of all the ash and small coals into the ash pan so that nothing falls on the cheese when you shove it under the fire in place of the ash pan on a home-made wire tray with a long handle. You have to wait till the cheese bubbles up and starts to brown, then you extract it and serve it up with the brown sauce. Mission accomplished.

The reason for the backup? There's always some rotten shit who either doesn't like you or your 'master' and when your rarebit is nearly ready goes past the stove and on the way gives it a good hard kick which disturbs the fire and drops ash and hot coals onto the cheese which you can eat yourself, negotiating the spoiling rubbish, after you've prepared the back-up, while keeping a wary eye out for the return of the saboteur.

Blister

As very junior first class apprentices we were subjected to abuse, intimidation and degradation, not by adults but by our own but more Senior Apprentices. Some of these seniors were more devious and cruel than usual. There were the ones, as a junior, you had to try and avoid like the plaque which wasn't easy.

There was one particularly notorious gang of half a dozen Seniors who carried out a reign of terror throughout the establishment. One of the things they organised was 'crow' hopping contests. They would collect four or five juniors and make them squat down and then force them to hop round like crows. It plays hell with the leg muscles and, depending how fit the victims are depends how long they last before excruciating cramp sets in and they collapse. The leg muscles ache for ages afterwards. The last one still going is let off, the others are given a good kicking.

Another one was beam hanging where they would force half a dozen juniors to hang from a dormitory rood truss bottom angle iron rail. As they dropped off through fatigue and aching fingers they were also given a good going over the last survivor being let off.

One of the worst atrocities I and a few other juniors had to witness was the humiliation and psychological cruelty of one of our own. The victim was a bit of a wimp but did not deserve the treatment he was subjected to. They had him tied face down on the dormitory table with his upper body bare. The table was positioned with the victim facing the pot bellied stove which had been stocked up till it was glowing.

A large dessert spoon was inserted between the bars of the front grate in full view of the poor sod on the table. Behind him out of his sight a pint enamel mug filled with iced water had a similar dessert spoon placed in it. The spoon in the stove, now red hot, was removed from the stove with a pair of tongs and shifted rearwards, out of sight of the victim, who was by now starting to panic. The stone cold spoon was removed from the mug and pressed on the back of the poor sod right between the shoulder blades, he screamed the house down in pain, the spoon was removed and a huge angry blister developed before our eyes, with the victim sobbing piteously.

A couple of us braver ones reported the incident to the authorities, and to cut a long story short a full enquiry was carried out, and the half dozen were dishonourably discharged along with three Leading Apprentices who condoned the practices, and the rest of the establishment breathed a bit easier. The victim of the last incident recovered fully and ended up none the worse for wear.

Hat Damage

The Royal Navy Apprenticeship in my day, was one of the finest technical training courses around. My day was from nineteen fifty three onwards.

The good thing about this apprenticeship was that you got to try all the different skills required for the different trades. One of the first courses was woodworking or 'wood butchering' which was what turned potential Shipwrights on. I quite enjoyed it but I was at a bit of an advantage because I'd attended a Technical Secondary School and part of the curriculum was woodworking and we got to make various items that were useful in the home, and you were allowed to take them home if they were good enough, and mine were.

But I digress, there was always one of the blokes on the course who wanted to wind either a friend or an enemy up, but one of the best I remember was a couple of twin brothers who were always trying to get one over on each other. One of them came up to me, I don't know which one it was, you could never tell the difference, and asked me to keep his brother talking and try and entice him away from his bench. It was no skin off my nose so I got nattering to him and asked him to come over to my bench to give his opinion on my efforts on a particular job. He gave me his advice which I thanked him for and we carried on with our own tasks.

Shortly afterwards the knocking off hooter sounded and we grabbed our hats to dash down to the Mess Hall. Our conned twin grabbed his hat and a loud ripping sound was heard and he had a ruined titfer. His brother had nailed it to the bench. Not only did he have to buy a new hat, he was also charged with misappropriation of personal property and also being improperly dressed. Fortunately he didn't connect my diversion with his brother's act of vandalism.

Whitewash

As a lowly first term apprentice at HMS *Fisgard* you were at everybody's beck and call, especially the Leading Apprentices, unaffectionately called 'Hooks', who were in charge of the dormitories.

Most of them were proper bastards who practised sprogging and intimidation on the first termers, and allowed their mates in the senior classes to do the same. We in Exmouth Division were very well treated in comparison the reason being that we had the best Hook in the whole Establishment. The reason for his relatively easy going, genial attitude to us was the fact that he was 'the' boxing champion of the whole establishment and could afford to be magnanimous. We had the happiest dorm of the whole camp and always won praise for cleanliness and tidiness by the inspecting Officers on their weekly tour of inspection.

The reason was that we all worked as a team led by our Leading Hand. Usually in the past I was in charge of window cleaning and dusting, but last week I was on floor polishing with a great big felt 'bumper' but this particular week in preparation for the inspection our reversed Hook put me on whitewashing the steps, both front and rear.

I'd already done the entrance steps from outside and was setting up to do the ones leading to the corridor and the blue rooms. Our Hook wanted to go through to the corridor and I was in his way so I moved to one side and jokingly threatened him with the loaded whitewash brush. Unfortunately as I moved I stumbled and saw a splash of white wash heading towards him. The next thing I remember was waking up with a very bad headache, he was apologetic but it was an instinctive reaction. I apologised and explained that it was a dreadful accident but I understood and deserved the consequences. Much later in our service careers we came across each other and had a good laugh about it.

Aluminium Bowl

As has been said before, as Junior Apprentices there were many trades to choose from, and to help you make your mind up Their Lordships in their infinite wisdom enabled you to have a go at all the skills required for all the trades.

One of those trades tried out was Copper Smithing which as the name implies practises making and forming all thing copper. Pipes big and small, fittings, plumbing and sheet work. But this also includes working with aluminium but more of that later.

One of the Copper Smithing exercises was to make a small round bowl with a flat bottom. A disc of sixteen gauge copper is issued to each apprentice and they are shown how, with a different formers, mallets and hammers, the shape is gradually formed. Unfortunately in working the copper it becomes hard and brittle. The cure for this is annealing, which is performed by heating it to dull read and dipping it in water to quench it. It can also be left to cool in air but this may leave a scale skin on it whereas dipping it tends to remove the scale.

Once annealed in this way it is again workable and can continue to be formed. The annealing procedure can be performed as many times as required. Once the proper shape is achieved it is planished. This involves gently but firmly tapping it all over with a planishing hammer which makes small flat indents all over the surface. If you produced a decent well shaped bowl you stamped it with your name and it was chromed and you were allowed to keep it. I used mine as an ashtray.

But I digress again – you also had to make an aluminium bowl of slightly larger dimensions so that some of the better ones could be used in the Mess Hall as sugar bowls.

The method of manufacture was exactly the same as for the copper bowl but with one notable and very critical exception. Aluminium also work hardens and has to be annealed to enable it to be re-workable, but the annealing process is different because if you overheat aluminium it doesn't glow to show it's done, it just melts.

The way to avoid this is to mark the piece of hardened aluminium with 'soap'. We called it pussers hard which was available to all for dhobying purposes, but any soap will do and all you do is put a large cross on it with the soap which doesn't show and heat it slowly and evenly, eventually the soap will char and show the cross in dark brown or black. The disc is then quenched and is annealed, easy eh!

But there's always one rotten sod waiting in the wings who gets his buddy to distract the unfortunate victim who is interrupted just after he's marked his job with the soap, and leaves it on the bench to attend to the bloke's problem. The rotter moves in with a damp cloth, rubs off the soap and returns the disc to the bench.

The victim returns, heats up his now unmarked job and it eventually just disappears before his eyes!

Twelve From Five

As a lowly apprentice, paid a pittance and half of that going on laundry, it was difficult to manage, especially if you were a smoker, but we learnt to be thrifty. One thing we didn't use was the canteen for extra food or drink, we would use the Mess Hall to thieve, sorry borrow, anything to supplement our diet. Bread was good as you could toast it on the dormitory stove and grease and jam were easy to smuggle out. Now and again one of our mates would get a food parcel from home and he would share it. We took turns sending begging letters saying how bad the food was, and it always paid dividends. Another way was that a few of us had distant relations living locally, and they were always good for a handout because they usually wanted rid of you.

Supplementing the diet was easy, it was fags we had a real problem with. Now and again one of us would get a Postal Order in the mail and we would splash out on woodbines from the canteen but that wasn't the end of it we also bought fag papers and filter tips.

The routine was to buy five Woodbines, because the canteen didn't sell rolling tobacco, and carefully slit the papers and place the tobacco in an air tight tin. A fag paper was extracted from the pack, a filter was placed at one end and a small amount of tobacco placed on the paper and rolled up the glue wetted and the thin fag sealed. The reason for the filter was not for health reasons, but so as not to waste tobacco. Doing it this way a dozen or more fags could be manufactured from five Woodbines. Then for one glorious moment we could smoke our heads off.

Abandoned Mug

Once again as junior Artificer Apprentices we had a difficult choice to make. What trade to adopt, because you are going to practise it for another three years as an apprentice and then the next twelve years in the fleet, a daunting thought. So the early days are spent sampling the different trades.

One of the first trades to try was woodwork, with a view to becoming Shipwrights. But of course being young and daft we used to now and again get up to mischief in one way or another. I have already related the scam of the damaged hat, and some of the devious goings on in the other trade trials and tribulations.

But the one I'm about to relate also happened in the joinery section. Midway through the forenoon and afternoon sessions we had a stand easy break and you could nip to the local canteen and get a mug of tea. The Navy thought of everything and you were issued with a pint enamel mug and you printed your name in large letters on its side so that it didn't get nicked, because some of our reprobate classmates would nick anything that wasn't nailed down, excuse the pun.

If you wanted to get your own back, for either a joke or revenge on somebody in your class, all you had to do was wait till stand easy when the pint pots were full of tea. Get an oppo to distract the mark away from his brew on some pretext or other. When he's out of sight, empty his mug into another spare, take a serrated nail, and with the aid of a punch, hammer the nail through the base of the mug and firmly into the wooden bench in its same exact spot. Then replace the tea into the now well secured mug. The beauty of a serrated nail is that it is easy to drive in, but almost impossible to pull out.

The mark returns to his well earned mug and nearly tears his finger off in his attempt to have a mouthful of tea. He also has to remove the damaged mug by brute force and also have tea all over the bench, and he also has to buy a new mug.

HMS *CALEDONIA*

Sunday Breakfast

Sprogging wasn't too bad just so long as you kept your head down and did the graft required as long as your 'master' didn't make too many demands. That is until the weekend when most of the seniors in our Division, and I'm sure it wasn't the only one, who demanded breakfast in bed on Sunday morning.

It was always the same, two slices of toast and a runny egg! The problem was basically twofold, firstly each individual was allowed only two slices of bread regardless, so you had to go round the buoy, which was cheating, and the civilian server attendants were eagle eyed so you had to be very careful. The trick on the second trip, you just asked for one slice of bread and a runny egg 'please'. You then got yourself back to your mess table where a reliable mate was guarding the first two slices. The third slice had its centre cut out to the size of the outskirts of the egg. The egg was then carefully slid from the plate onto one of the whole bread slices, the cut-out slice was then laid carefully over the egg to frame it and the other intact slice carefully laid on top to enclose the egg. The second problem now reared its ugly head, removing the triple decker wedge from the Mess Hall as removal of food from same was strictly forbidden.

The only way to do it was to carefully stick it under your armpit without crushing it or letting it slip, which made for rather difficult walking and could be a bit of a giveaway if you were not very careful.

Once you're clear of the Mess Hall you're just about home and dry, that is unless your encounter an enemy of your 'master' who, suspecting that something might be a bit naughty hails you like a long lost friend and slaps you on the upper arms both at the same time! Result, one badly squashed egg that is now running down your shirt and you have to try and start all over again if you're not re-spotted in the Mess Hall.

If you manage to evade all encounters and achieve your 'master's' abode you still have to toast his bread on a one bar electric fire without setting light to it because it has grease (margarine) on one side of each slice. Meanwhile you have to attempt to keep the egg at least lukewarm, and still runny!

The bonus was that you got to eat the middle of the third slice if you were lucky.

Taffy Inon

HMS *Caledonia* above Rosyth Dockyard in the Kingdom of Fife was the specialist training establishment for Engine Room and Shipwright apprentices after our initial training and selection at HMS *Fisgard*.

Coming from being top of the heap, Senior Apprentices, we all ended up as the lowest of the low again, and ended up sprogging again.

My 'master', Taffy Inon, was an evil little Welsh git who nobody liked, not even his own classmates, as he had a vitriolic temper, and no personality whatsoever. This particular Saturday lunch time he decided he was going for a run ashore so he instructed me to clean and polish his shoes, with my own gear of course, because he had nothing like that. The snag was that I had just dhobied out my polishing duster, but I found a scruffy looking lump of rag in his locker that would be adequate for the job.

I was giving his footwear the final buff up with this rag when he suddenly appeared and went absolutely ballistic.

'What the f—in' h— do you think you're doing with my spare pair of kegs you imbecile?' he screamed.

All his dormitory inmates stopped what they were doing and swivelled round to see what all the fuss was about and realised what I'd done wrong, and as soon as they realised what it was, as one they all burst out into uncontrollable laughter. So he snatched his shoes off me, shoved them on his feet and rapidly disappeared, very red faced.

But it had its good side, he eased up on me a lot after that incident!

Knock Off

I ended up a sprog again, in HMS *Caledonia*. From being a Senior Apprentice down in *Fisgard* to the lowest of the low again with the same types of intimidation, but even worse.

Some of my classmates in other Divisions still had their bags packed and stashed under their beds ready to do a runner if things got too bad, and they were definitely heading that way.

The incident that made it more bearable for me had a very innocuous start. I'd been to the Sick Bay suffering from what I thought was a heavy cold and all I wanted was a couple of aspirin, but the Quack insisted that he looked me over and eventually diagnosed flu and dosed me up and told me, as it was the weekend, to keep dosing up and turn in as much as possible, so I did just that with a nice empty dorm. That is until three Senior Apprentice louts started to terrorise me.

They came through the joint looking for trouble and one of them saw me in bed even though I tried to hide, and wandered over and asked what was wrong. I explained that I had the flu and the Doctor had dosed me up and prescribed bed rest. Then he called his mates over and discussed my case with them then enquired of his mates if my counterpane was upside down? The other two agreed that it was and ripped off all my bedding and went on their way laughing their heads off.

I climbed out, feeling not too good, made up my bed and turned in. Half an hour later they came back and the same thing happened again, and again as they sodded off I climbed out made up the bed again, feeling a little bit worse, and got in again.

They came back yet again and they'd obviously been in the canteen for a few beers, and the same thing happened yet again, and I was beginning to get a bit teed off with the whole charade, the counterpane was wrong way out again, all my bedding was on the deck again, and again they went away laughing, so again I climbed out and made the bed up again and was about to turn in, by now bubbling inside, when the ring leader returned and was about to rip my bed up yet again, so I asked him not to do it, and he asked me in a sneering tone of voice what I was going to do about it if he did. I didn't reply but just shrugged my shoulders, so he grabbed the counterpane and I hit him!

It was a lucky blow 'cos I got him in the solar plexus and obviously winded him, and he went down like a sack of spuds. Unfortunately, on the way down he cracked his skull on the cast iron bed head with a horrible crunching sound and he lay there bleeding profusely. What to do? On hearing his drunken mates coming back, I turned in, and feigned sleep.

They came across their mate and discussed, in a drunken manner, what might have happened, and one said that I must have hit him a bit hard and they lifted him up and he staggered between them never to be seen near me again and I found a little bit of respect from the seniors and they didn't trouble me much anymore thanks to a cast iron bed head!!

What's A Beefer (Baines)

Being the lowest of the low as new larkers at *Caledonia* was, to put it bluntly, miserable. You were at everybody's beck and call, especially the top class.

One of the worst experiences you can be subjected to is when Seniors who have had a skinful ashore come back aboard and head for a junior dormitory and turn all the inhabitants out of bed for interrogation and trash all the bedding just for good measure, then get some poor unfortunates crow-hopping or beam-hanging or just being asked stupid questions. Sometimes an individual would be picked out and interrogated which on this particular night happened to one of our classmates. He was a good hand called Baines, Bunty to his mates, they had him stood on the table.

At this point I have to give a little explanation, in Naval slang a Beefer is a homosexual!

So they had Bunty on the table asking him stupid drunken questions and one of the interrogators came out with this one:

'What's a Beefer Baines?'

Reply:

'Making Honey Senior Apprentice!'

To which they all rolled about laughing and left leaving us to repair our beds and get some well reserved thanks to Bunty Baines.

That was one of the more amusing incidents, but generally it was very difficult to do your best academically and manually as well as coping with stupid drunken Seniors.

Cannon Fight

It all started innocuously enough with a couple of mates of mine experimenting to see how far they could propel a small ball of silver paper, or aluminium foil.

The method they initially adopted was to use of all things a locker key, the sort that has the shank drilled out to accommodate a pintle riveted to the base of the lock mechanism.

First you crush heads – the safety match variety, the red or blue heads tend to self-ignite! The resultant powder you pack into the hollow in the key, best held vertically. You then roll up a piece of aluminium foil as tight as possible and fit it on top of the match powder and tamp it down a bit to make it a good fit. Next hold the key in an aiming position, preferably slightly above horizontal, and secure, a portable bench vice would be ideal for this, then apply a naked flame to the shank of the key, a lighter or candle would be better than matches as you may run out of the latter.

Depending on the intensity of the applied flame, the crushed match heads become hot and eventually self-ignite and if tightly enough packed tend to explode which ejects the foil pellet, and a measurement can be made to check the distance fired. All very innocuous you say, but it started to escalate as someone found a larger hollow key so a bigger charge could be used and a bigger wad of foil used, and a competition was on to see who could fire the furthest.

New of this new competition spread and a lot more blokes began to be involved, then the workshop got involved and odd bits of bar went missing but not before holes of varying diameters and depths were added. Then some ingenious fellow drilled a small hole at ninety degrees to the main hole at its base and the touch hole was formed negating the use of the live flame for firing. This in fact now was a miniature cannon and more and more were being produced some even made to look like miniature cannons, complete with carriage, some even bored them out to fit ball bearings.

Eventually it had to happen, someone caused a misfire which narrowly missed another crew, who took revenge and fired back and also narrowly missed.

Then it escalated and barriers were erected at each end of a dormitory with bedsteads laid in a line on their sides with mattresses in front, and firing was shot for shot. It was great sport and every dorm in every Division got up for the game at the weekends, you've heard the saying about idle hands, but unfortunately an accident was bound to happen.

There had been a few minor injuries, and burns, but eventually the inevitable happened and one silly sod stood up to protest and the opposition had already fired the touch hole, and he caught it in his left eye. Fortunately for him the missile velocity was insufficient to penetrate his brain but he lost the sight of that eye and was therefore invalided out of the RN.

The repercussions were severe and the practice was terminated immediately and anyone found to continue the practice would be awarded a dishonourable discharge!!

Torpedo

Some blokes had hobbies, and most blokes had normal interests, but my mate Chippy Wadge had a more unusual hobby, he liked explosives!

His nickname was Chippy, which is the usual name for a Shipwright, but he was a fitter and turner apprentice the same as me, even he didn't know why he got the nickname. He had been slightly involved in the Cannon affair, but his main preoccupation was torpedoes and he co-opted my assistance to develop the design with a view to blowing up the ice on the static water tank as it was midwinter.

The mark two version torpedo was still the same cigar tube with the screw cap end, but the detonation system was redesigned. The explosive was still crushed up safety match heads, but the detonation for same was a one point five volt torch lamp with the glass envelope broken to bare the element which was then delicately wrapped in tissue paper with finely ground match heads contained within it. The lamp was connected to the battery via a piece of watch spring that was insulated from the aluminium tube and strapped to the outside of it. The other end of the spring is adjusted so as to not quite touch the rounded end of the tube. A wad is inserted in the tube to keep the battery assembly firmly in place. The remaining room is filled with Jetex pellets and finished with a coil of

Jetex fuse. Jetex was very popular in those days for propelling all sorts of models as when lit it produced volumes of gas which when contained with a small discharge orifice produced thrust. The screw cap of the tub was drilled in the centre to s sixteenth of an inch diameter. The end of the Jetex fuse was fed through this hole and the cap screwed tightly onto the tube.

It's supposed to work thus. The Jetex fuse is lit and the Jetex ignites, a good exhaust stream is established and the torpedo is launched in the static water tank. It speeds away and hits the far side where the spring connects to the tube dome completing the circuit, the lamp element illuminates, burns out and fires the fine powder surrounding it which fires the main charge, mission accomplished.

We had our redesigned torpedo ready to go but Chippy wasn't happy with the size of the Jetex hole in the cap so he decided that we should test it in the blue room because there was a huge pot dhobying sink, oblong, low level, ideal for our test. To make it safe a piece of insulation was stuck to the dome of the tube to prevent ignition and we would be able to recharge the Jetex after the trial.

So the sink was filled to the brim with water. The Jetex fuse lit and the pellets ignited well and Chippy put the torpedo in the water and let go. The thing set off like a good 'un but we both saw the insulation fall off the nose. Chippy was nearest and he grabbed for it but it was just a bit too late. It hit the outer corner of the sink and exploded as Chippy grabbed it and it blew up his fingers a bit, and then the corner of the sink broke off and about ten pounds of pot hit his foot, followed by gallons of water.

There he was for all the world doing an Irish jig with his burnt hand flapping about and hopping about on one foot.

I couldn't help myself, I just burst out laughing, and Chippy was not amused, but his injuries were superficial and he soon recovered. Not so the sink, the soot was washed off the damaged bits and reported. After an enquiry it was determined that it was very old and its failure put down to fair wear and tear.

We went on to develop the mark three and eventually blew up the ice on the static water tank in a very spectacular fashion in front of a very appreciative audience.

Teddy Boys (War)

I and my classmates, having enjoyed *Fisgard* as Senior Apprentices, were now at HMS *Caledonia*, the lowest of the low again. We were again at the beck and call of the top class again, and they didn't take prisoners. Some of our classmates didn't even bother to unpack, with a view to going over the fence and deserting, it was that bad. But somehow most of us survived relatively unscathed to carry on with the training.

HMS *Caledonia* was situated above Rosyth Dockyard, the nearest decent run ashore being Dunfermline, a nice place with lots of history and plenty to see and do, but a bit hilly in places.

Eventually our class had progressed past the junior stage and were at the point halfway through our apprenticeship where we didn't worry anyone and nobody worried us, a nice and easy state to be in so that we could concentrate all our efforts on the practical side and also the academic side which I was progressing a lot better with, but a trip into Dunfermline always helped to ease the tensions of work. The only snag was that we had to go ashore in uniform which made us a bit conspicuous but at least you knew who your shipmates were.

We heard on the grapevine that a worrying trend was developing down south and rapidly spreading northwards. It was the Teddy Boy era, a load of teenage yobs dressed in drainpipe trousers, winklepicker shoes and stupid haircuts, roaming the streets mob-handed causing mayhem with the general public, and especially decent teenagers.

This cult spread rapidly north eventually crossing the border, surfacing in Scotland and inevitably Fife and of course Dunfermline. Of course the main targets were decent teenagers especially the English and easily recognisable were us apprentice Tiffs because of the uniform. The louts' favourite ambush site was the lower Bus Station where they knew our lads had to catch the last bus of the night. If they missed it they would be in the rattle for being adrift. Some of the lads turned up at the guard house very late in terrible states, covered in blood and bruises with their number one uniforms trashed. This for us was the final straw, disrespecting the uniform, enough was enough and the seniors, Hooks included, started recruiting forces and assembling arms.

Fire extinguisher hoses went missing to appear in the coppersmiths shop where they were filled with lead and distributed to the lads. Socks loaded with sand and tied off, leather gloves complete with homemade knuckledusters, welding gauntlets loaded with palmed steel bars.

Our authorities aboard outlawed vandalism of the fire extinguishers but turned a blind eye to other activities, including being adrift.

The operation on consecutive evenings was planned with true Naval precision. Those seniors who owned cars loaded them up with troops and went to Dumps in convoy, others filled the local bus service and all converged on the lower Bus Station, with perfect timing.

It just so happened that the Teds were out in force that first evening and night and they got a good hiding, and the local law turned a blind eye to the whole rout because they too were being abused and outnumbered by those thugs.

The second night the Teds tried to regroup and recruited a few extras, but they still weren't good enough and got a good hiding again.

The third night the Bus Station was quiet and running smoothly as normal with no sign of the opposition, so we all wandered up into town and spread out to see if we could winkle out a few of the opposition but no such luck, they had all gone to ground and we all went back aboard feeling quite elated at a job well done. We even got a veiled pat on the back from the local press.

Danny Kay (Punch-Up)

At *Caledonia* a few classes above me was a lad called Stanley Kay, Danny to his friends. I didn't know him very well at the time of the incident but we all knew him by reputation. He wasn't particularly tall but what some would call well built, but to us, his friends, he was built like a brick shithouse (sorry, toilet).

After the Teddy Boy phase there was an occasional isolated incident of actual or attempted assault so the lads used to go ashore to Dunfermline in small groups, although some preferred to be, as Danny, a loner.

It would appear that the Dumps yobs had noticed this, and Danny was a creature of habit and his runs ashore were always the same routine. He was a motorbike buff who used to go for long burn ups on the crap

Scottish roads on an old Triumph Thunderbird with a knackered rear spring hub which made it a bit wobbly at speed, his price and joy. But his routine, always on one of his bike trips was to stop off at his favourite chippy for his usual pie and chips with mushy peas and plenty of gravy. His routine had been noticed by a group of ex-Teddy Boy yobs who wanted a large measure of revenge.

There was Danny sat side saddle on his Thunderbird devouring his supper when half a dozen locals surrounded him with a view to administering GBH but Danny was not about to be intimidated and carried on eating his supper until one of the opposition kicked his beloved motorbike.

A witness, the lass serving in the chippy, described the action as quick and efficient and called the ambulance service to administer to the injured who ended up being hospitalised suffering from various fractures and dislocations. Two of the gang fled almost unscathed but Danny later identified them in a Police line up and all of them were convicted of affray and severely dealt with.

The witness also explained to the Police that after the altercation Danny finished his supper, kicked off the Thunderbird and rode off with a cheery wave, departing the scene of mayhem. The victims did not know that their adversary was a black belt karate expert.

I and my mate came across Danny later and we ended up good friends with some good, but no great, tales to tell.

Paddy's Rugby

One of the nicest and most likeable blokes I ever met in the apprentices was a bloke called Michael Ryan, but Paddy to his friends. He was hail fellow well met, always a kind word, never an ill one. Hell of a sense of humour and built like a brick toilet to boot. You could take the urine out of him, try and wind him up but to no avail, he would just grin at you and slap you on the back.

But, get him on the Rugby field and he changed into an animal, a wild beast, a monster! His position was prop and if there was any skulduggery in the scrum you could be sure Paddy was the perpetrator who if challenged by the Referee would look the picture of angelic innocence until the next time.

The next match he had was against the Army and Paddy was playing his usual brutal game, but they Army lads knew him of old and were determined to subdue him one way or another. Eventually a massive ruck evolved with a lot of writhing, grunting and groaning, and when the Referee's whistle blew and bodies started to untangle themselves, right at the very bottom was Paddy out cold, with his left arm looking somewhat poorly positioned. He didn't respond to stimulation on the side line so a blood wagon hoisted him off to Sick Bay and the game proceeded with the Army eventually triumphant.

A few of us wandered down to Sick Bay to see how Paddy was doing and were told that he'd been shipped off to hospital in Dunfermline so I rang up and found the time for evening visiting and a few of us went ashore at the appropriate time to see how he was doing.

It was only two visitors at a time so we spelled it between us. When I got to see him he was almost his old self, laughing and cracking jokes as usual, but I sensed some underlying trouble within him. During a change-over of visitors we were alone for a little while so I whispered in his ear and asked him what was up, and he very seriously said he was worried about his arm not working! But then somebody else rolled up and he was back to being Paddy again.

He was eventually discharged back to us after recovering from his mild concussion but he had his arm strapped up in a frame which kept it ninety degrees from his body and elbow bent at ninety degrees, as if to make a Nazi salute. He even had to turn in with his rig still attached. He also has to keep going back to hospital for physiotherapy but it didn't seem to be doing much good.

Eventually the hospital and the Navy reluctantly admitted that nothing more could be done for poor old Paddy and he was invalided out of the Navy and was given a lowly job in Rosyth Dockyard. But despite all the trauma and disappointment he still stayed the cheerful Paddy we all knew and loved.

Boiler Makers

Housed in the same area as the Engine Smiths where the Boiler Makers, which is a bit of a misnomer really because their usual work was boiler repair.

In the era I'm talking about the most popular and prolific type of boiler was the Admiralty Three Drum type in a triangular configuration. The top of the triangle was the larger diameter steam drum complete with water level controls, gauge glasses and steam outlet main valve.

This drum was connected to two smaller water drums at the base of the triangle by two distinct rows of tubes. The outer rows were the down comers which brought hot water down from the steam drum, and the up risers or generator tubes as the inside of the triangle was the furnace where the fire heated the water to generate steam on its way up to the steam drum.

The Boiler Maker practised removing and replacing tubes and also making what was called joggle patches and riveting them on to the drum to effect a repair, and when they were riveting it made a hell of a racket and some poor sod had to be inside to hold a dolly against the head of the red hot rivet while outside they formed the head of the other end. The one inside has to be reasonably small to get through the manhole, and once inside has to wear really good ear defenders.

All these Boiler Makers need is the smallest excuse, like a birthday or someone they don't like, and in the water drum they go, the lid is closed on the victim and away they go, the whole gang attack with assorted sledges and large hammers.

By the time the victim is let out he's reeling and stone deaf. It's what's called rough justice. Or maybe a riveting experience!

Engine Smiths

The *Caledonia* workshop was a hell of a place and catered for all engineering trades. Apart from us Fitters and Turners there were Engine Smiths, Copper Smiths, Boiler Makers, Pattern Makers and Foundry Men, the later being recruited from Civvy Street, and employed on Depot Ships only.

Of the remaining three trades the most dangerous to us Fitters and Turners were the Engine Smiths, the reason being that we had to go to their section to acquire the material to make our lathe tools. This material came in square section suitably sized to fit our lathe tool posts, but in ten-foot-plus lengths, and instead of cutting it off to the length we required with a hacksaw, or let us do it with the same, they insisted on heating it till 'cherry red' and then cutting it off with what they called a cold set. This involved a horizontal blade mounted on an anvil with its square shank fitting snugly into a square tapered hole in the blunt end of the anvil. The almost white hot bar is laid at right angles onto the mounted blade with the required length overhanging. Then the one in charge places another blade on the red hot bar above the lower blade by means of a long wire handle. The striker then comes along with a sledge hammer and batters the top blade onto the bottom blade, so cutting through it like a hot knife through butter.

This pair of Smiths were supposed to finish the cut very gently, but no, the last blow is the hardest of all and suddenly if you've not already been warned you're dodging flying lumps of red hot metal, much to their delight.

Welding

Welding was something we all tried, but in those days only two types existed. The gas type involved bottle gas, a live flame and filler rod to match the materials being welded together. The other method was arc welding which uses an electric spark at very high temperature produced from a transformer with the power going down to a positive lead to an electrode of the right material with a flux coating to prevent oxidation, the electrode being held in a spring clamp handle. Another lead, the earth, is connected to the material to be welded and as soon as electrode touches the material a very hot arc is struck which melts the electrode and when in the right position, welds the two pieces together, the flux also melts and covers the weld bead to prevent premature cooling and cracking and also as an antioxidant.

The object of this part of the apprenticeship was to show that welding was possible and in some cases essential for fabrication and repair. Arc welding in particular is difficult to set up, the variables being thickness of material to be welded, material to be welded, material type, amperage required, electrode diameter, and flux type to name but a few. One thing to avoid is arc eye which is caused when a struck arc is directly eyeballed without an intervening dark screen. The problem comes not there and then, but late at night or early morning when suddenly the victim wakes up with terribly irritating eyes that cannot be treated, but has to be endured and lived through.

But I digress. The welding section was split into steel plate bays or booths, self-contained with a power supply, welding plant, bench, vice, lighting and all the tools and bits and pieces to be able to practise arc welding. These booths were portable so even the deck was steel plate. This is where the snag occurs, or opportunity for mischief, depending which side you're on.

The trick is as always to keep the victim still and occupied by one of the team while the other ensures that the welding plant next door is set to the right amperage and the earth clamp is attached electrically to the marks booth, next have a handheld screen ready, then a quick peer round the corner to check that the mark is in the correct position, then pounce and a quick arm strike between the deck place and the boot heel plate and a quick stitch, if time do the other boot.

As in all good books the hero is welded to the spot. In this case literally!!

Swarf War

Our days of hand fitting were over, and we had escalated to the machine shop. Row upon row of belt driven lathes, I couldn't wait to get my hands on one. My mate was the same as me, keen as mustard. But there are those who were either indifferent or downright not at all interested. That was the case of my mates towny Dennis. He was the sort of bloke who'd talk the hind leg of a donkey but say nothing much of interest of help. His passion was betting, wagering, gambling or running a book on anything you care to name.

Before you start using a lathe you have to make your own lathe tools by shaping, hardening, tempering and sharpening them, and how well you made them determined how long they would last in use.

We all started the first turning exercise by centre drilling a bar and running it between centres and driving it in the lathe with a carrier which is driven with a peg on a face place of the lathe spindle. We were then instructed to turn the bar down to a predetermined diameter. Most of us had formed a chip breaker edge on our turning tool to make sure the swarf formed small bits which were contained in the lathe tray. But not Dennis, he just hadn't bothered with the chip breaker edge, it was just too much trouble, so what happened was that instead of the swarf breaking into bits, his came off in a continuous stream, wriggling its way to the tray, then wriggling off the tray and onto the deck, and started setting off in a southerly direction. So Dennis, quick as a flash, opened a book on how long the swarf would grow before it broke off, and made chalk marks on the deck in way of its southerly course. Believe it or not there were plenty of punters and soon blokes were cheering and shouting and a few were contemplating sabotage because the wriggling ribbon had overtaken their chosen mark. By now the racket was so loud the Regulating Chief Stoker came out of his office to see what the fuss was about and even the instructors were getting more than a little excited as the thing kept on its wiggly way. The RCS shook his head, gave up and went back into the office for the quiet life, and if asked knew nothing about it.

All eyes were on the swarf snake and nobody checked the lathe, and at the 35 foot 6¼ inches length, after a lot of money had been lost, the cutting tool hit the carrier, the tool cutting edge was badly damaged and the swarf snake broke. Amen.

One guy made a lot of money, and Dennis was also well pleased and sod the turning job!!

Morris 12

I was middling senior in the apprentice hierarchy, where nobody bothered me, so I bothered nobody either. I was doing OK down the main factory because I'd found something I was relatively good at.

I'd also established quite a rapport with the Regulating Chief Stoker who was reputedly a bit of a stickler. The reason for this was that I'd helped him out a couple of times with his car engine as he was a mechanical no no when it came to cars. He had a Morris 12 that when I first looked at it sounded like a tank so I adjusted the tappets which quietened it down more than a bit. The other problem was that it was a very bad starter and on investigation I found that it wasn't the engine it was the starter motor. All I had to do was strip the rotor out of it, put it between centres and skim up the commutator and polish it, then reassemble it and fit new brushes, having checked the bearings. It was as good as new with no more starting problems.

A couple of months later he came back to me to ask if I wanted to buy the Morris as he was looking for a newer car and I said I would if the price was right so we negotiated and struck a good deal as far as I was concerned.

So I took over my new acquisition and drove it round 'Caly' to show it off but it was running badly and sounded rough so, come the weekend I decided to investigate, did a compression check and got a very poor result so decided to do a top overhaul, and discovered that all four exhaust valves were burnt and leaking to varying degrees. There was a Morris main agent in 'Dumps' who supplied all the bits so the next weekend I had it all back together by Sunday evening. It was now running a treat so I decided to brag a bit and drove at dusk from the top sheds down the hill and right towards the factory when suddenly there was a double bang and I shuddered to a halt. I leapt out to find out what I'd hit and found myself surrounded by the Trailer Pumps crew who it transpired were doing hydrant checks for exercise. In the dusk I hadn't seen them and they hadn't seen me and neither of us had any lights on! They had parked the Trailer Pump with its wheels in the gutter and the towbar poking into the road. I'd come along and hit the towbar with my nearside wing, spun the Pump right round, missing all the crew, and punching a hole in my rear nearside door, which left the lot of us feeling more than a little bit foolish. It was all hushed up as nobody was hurt, the Trailer Pump suffered no damage, and I was still in one piece. But my poor car! So I took it back up the hill and spent about a month of weekends panel beating and gobbing up until, although I say it myself, it came out a pretty good job.

Paint was the next item on the agenda. I decided on a two-tone effect, maroon bottom half and beige top half. I used brush on Valspar and again, bragging, I reckoned it came out looking pretty good.

I again trotted it down, carefully this time, to show the Chief Stoker who, I have to say, was very impressed!

Then came upon us the Suez Crisis, and the bonus for me was that learner drivers could drive unaccompanied by a qualified driver while petrol rationing was in force.

I must have travelled most of Fife on my tod 'cos I had a racket going which netted me plenty of petrol coupons, and I had plenty of money 'cos my money lending racket was proving very lucrative! I had befriended a local family, an ex-miner and his wife, and I used to take them on tours of the area, all of which honed my driving skills ready for the day when the crisis would be over and I would have to take the dreaded driving test.

Driving Test

The Suez Canal Crisis was a god send to me because, having the Morris 12 as a learner driver I could drive on my own with no qualified driver alongside. But the writing was on the wall, the crisis was coming to an end so I decided to slap in for my test in Dunfermline as soon as possible, and it came as quick as that.

I was quite confident in myself, my only fear being the hill start as the handbrake was a bit iffy and if I applied it well I had a real problem getting it off, and some of the hills in Dumps were more than steep.

Come the day of the test the examiner climbed into the passenger seat, commented on the nice condition of the car, and then went all po-faced at me. We went in and around Dumps and I was quite happy and confident, that is until he directed my to below the lower Bus Station and then I knew that this was the big test as the only way from there was up a very steep hill to the main drag. The usual routine was as I expected, he said, 'When I strike the dash with my folder you execute an emergency stop,' and then *bang!* and I anchored up with a huge heave on the hand brake and it held! We had a little chat about rules of the road and then I was instructed to carry on, and now it was do or die time. I was in first gear the clutch biting, the revs just right but my left hand couldn't shift the brake lever up to clear the ratchet. There was a bit of a scuffle going on outside so I just nodded to the left and mentioned it to him, he glanced out, and my chance came, my right had covered my left and an almighty heave and we were away again and I recovered the steering wheel with my right hand before he turned round again.

Happy to say I and the Morris passed and we both went on to do great things together.

Summer leave was coming up soon so I decided to take the car down home and show it off. The snag was the steering was a bit dodgy with quite a lot of free play so I purchased a set of king pins and bushes for the job. That is where I made my first mistake, I didn't buy the reamer to go with the sets. I didn't know about reaming the bushes after fitting them to the yokes. So I went blindly on and eventually managed to get them back together by hand scraping the bushes and emerying them smooth, that was my second mistake. The steering was now quite positive after I'd removed a couple of shims from the steering rack input shaft to take up wear, and all seemed A-OK.

So off I went home on leave, bags packed, tank full, etc etc. Of course in those days there were no motorways in the north and very few dual carriageways, just the old A1, with all its twists and turns. The further I got down South the worse the steering got, not slacker but stiffer. The mistake I had made was the emery I had used. Some of the abrasive was still embedded in the white metal of the bushes and was slowly seizing it up with every steering movement. I thankfully arrived home, thoroughly exhausted because for the last hundred or so miles I hadn't so much steered the dammed thing as 'aimed' it as it would only shift in jerks, mostly overdone. What I at last arrived home the steering could be officially classed as seized solid.

The old Morris did me proud for a long, breakdown-free time as I did the king pins properly and started to teach my elder brother to drive. Eventually when I went out into the fleet and got a draft to the Persian Gulf I donated the old girl to my brother, who enjoyed it for a very long time.

Bubble Car

The mid to late 50s saw the arrival of all sorts of bubble cars. All sorts of shapes and configurations but most of them were three wheelers. There was the line-ahead version with a canopy like a fighter aircraft which hinged open sideways, the Messerschmitt KR175. Then there was the rather wider version which I cannot remember the name of where the whole of the front hinged vertically open taking the steering wheel with it.

My mate and I were still trying to figure out how to accurately adjust the ignition timing on a two-stroke motorbike when one of our classmates drove up in one to brag about how good it was and offered us a ride in it as a passenger one at a time. We declined his less than generous offer so shrugging his shoulders as much as to say your loss he climbed in and slammed the front door shut which brought the steering wheel back into position.

My mate gave me the wink and crouching low we snuck round the back of the vehicle where the driver couldn't see us and we gently and slowly lifted the rear wheel off the deck, as it was rear wheel drive. He started the two-stroke engine stuck it in gear and eased off the clutch. We could feel and hear the drive wheel turning. Our intrepid driver thinking he'd lost first gear changed up into second and gunned it to no avail so he went for top gear, third, and really gave it some welly.

Unfortunately when he at first rolled up he was facing the open compound of the full coke store. So when he got to full chat we dropped him and with a screech of tyre and us covered in acrid blue smoke he shot off like a rocket straight into the coke pile and almost completely buried himself while me and my mate were still tinkering with the timing problem. We raced to assist him, asking what happened but he was so bewildered he couldn't even speak. It took us a monumental effort to keep a straight concerned face. To this day he doesn't know exactly what happened but I think he has his suspicions but he always was a bit thick!

Steaming Bike

Being an apprentice fitter and turner I liked to keep my hand in during my spare time and I was tinkering with a mates two-stroke BSA Bantam which had a nasty habit of trying to break your leg when you tried to kick it off.

I knew it was the ignition timing but it was a bit of a b— to adjust correctly or accurately. I have know of two-strokes starting up backwards without the rider knowing until he lets the clutch out and finds he's going the wrong way. Indeed I know of a certain bubble car manufacturer who has exploited the phenomenon to move a lever to alter the ignition to give a reverse gear.

But I digress, I was busily squirrelling away when suddenly there was a roar of a motor bike and another mate of mine came steaming in full tilt, hit a big concrete parking block and shifted it a couple of feet put the bike of its side stand, leapt off, threw his gloves, helmet, goggles and jacket at the back of a load of assorted rubbish and knelt down beside me and whispered, 'I'm helping you OK,' so I just nodded and carried on with what I was doing. A minute or so later the doorway again darkened and I looked up to see an enormous 'Plod' in biker gear who said in a deep rumbling voice 'Has anybody just come in here?' We both shook our heads, then I had an inspiration and asked him if he knew anything about two-stroke timing adjustment, to which he replied in the negative and he went on his way. As the doorway once again became illuminated we were aware of a cloud of smoke and steam rising from my mate's still hot bike which slowly started to fill the shed. It was lucky he'd been such a big sod otherwise he would have seen it for himself.

It appears that my mate was running late for a date and had shot past 'Plod' at a rate of knots as he was parked on the kerb edge, so realising what had happened he decided to go for broke and just made it in time.

Coventry Climax

The proceeds of the bike sale and my mate matching it bought us an old sports car that an RN Instructor wanted rid of when he got drafted.

It was an open four seat tourer which had been a rag top but had only the folding frame left. In its day it must have been quite some bus as although the upholstery was quite well worn it was leather throughout. The engine was a 1.5 litre overhead cam Coventry Climax which was very much in need of a top overhaul as the exhaust valves were leaking like sieves, as was shown by compression tests. We lifted the cylinder head and our diagnosis was proved correct as all four exhaust valves were badly burnt but the seats were almost near perfect and would easily lap in.

We thought we might have trouble locating a new head gasket as well as four new exhaust valves, but not a bit of it. There was a Coventry Climax agent in Edinburgh and a local garage in Dunfermline was our go-between, and before we knew it we had all the gear we needed to complete a full repair. With a new battery and the commutators of both the starter motor and dynamo turned up to perfection and new brushes fitted to both, she flashed up first crack, and sounded like a sewing machine.

We decided to take her for a spin the very next weekend and invited a couple of classmates to join us for a trip to Alloa and back. I drove out and we stopped at a local Hostelry for lunch and a couple of beers, then when well satisfied my mate took over the helm and proceeded to drive us back. He was keen to show our guests how pokey the motor was and took a rather sharp left hand bend far too fast and we ended up teetering on two offside wheels before going over on its side and tipping us all out into the middle of the opposite roadway which had just been resurfaced with granite chippings. It was fortunate that nothing was coming in the opposite direction and it was doubly fortunate that is didn't roll right over or we would all have been trapped. As it was the old bus just slid along on its side with a terrible tearing sound!

One or our guests unfortunately suffered a bad head injury but the rest of us just suffered rather nasty gravel rash and all of us ended up in Alloa Hospital for observation. The ward we were in was full of old fogies who were mostly suffering from bladder trouble caused mainly by not drinking enough. So we vowed and declared we would not suffer the same fate!

I think we were a bit of a breath of fresh air to the younger of the female nursing staff and we had some good laughs, but all good things must come to an end and three of us were discharged and had to get the bus back to Dunfermline. We all looked a mess as our uniforms had suffered gravel rash too and I'd lost my hat as well.

We got on the bus and the only seats available were right at the back and of course well to the rear of the rear axle, and the first sharp corner we took we all thought we were goners again, a very strange sensation when you've experienced the results of a previous one.

We got the bus from Dumps to 'Cale' and wandered into the Guard House and were immediately put in the rattle for being AWOL and being improperly dressed until we explained to the duty OOW that we had been involved in a car accident and that one of our number was still in dock.

After a lot of to-ing and fro-ing we were released and returned to our dorms with more than a little urine extracting. The real cost though was the expense of acquiring new No1s and in my case a hat! The good news was that the fourth member of our party returned a couple of days later none the worse for wear.

Matchless

The Senior Apprentice class was leaving to join the Fleet and as usual a few of them had bits and pieces they wanted rid of, but this particular bloke wanted rid of a 250cc Matchless motorbike at a ridiculously low price.

It was an offer I couldn't refuse 'cos it was in good nick and sounded like a dream, and of course came with all the gear, boots, leathers, helmet and good goggles.

Me and my mate, who also had a bike, used to regularly do tours of Fife, quite often the coast road that could be quite spectacular especially if the weather was a bit blowy.

Sometimes we used to go to the Ochill Hills which, when we had reached a certain altitude, the view was spectacular, and on the way back we'd stop at a booze on Loch Leven side and have a couple of halves before going back aboard.

The only problem I had with the bike was that for a 250 it was very heavy, not that it mattered when I was rolling, it was when I stopped that it became apparent. I'm not all that big and strength I haven't got a lot of, in fact I'm the original seven stone weakling, but I've evolved ways of overcoming these deficiencies in all sorts of ways, but with the bike if I didn't stop exactly upright I'd fall over, then I'd have trouble getting the bloody thing upright again and if I over did it a bit it would go over the other way, and I'd have the same problem again.

You look a bit stupid if it happens say at traffic lights, or a compulsory stop junction. People tend to laugh at you, so eventually I reluctantly gave up biking for good, but at least I got a good price for it and the gear and made a tidy profit and my mate used to take me to our same haunts on his pillion and we enjoyed the ride just the same. In fact later on we met up again when he had taken up road racing on an old but good Manx Norton and I ended up for a while being his mechanic and pit manager which took us all over the country on a Triumph Thunderbird and sidecar chassis with the Norton on the chassis and me on the pillion again.

The Test Job

At a recent Artificer Apprentice reunion I was reminded of a favour our class did for a Kiwi classmate. This particular bloke was a good hand, over six foot tall, and built like a brick toilet, a good boxer, an excellent rugby player and an absolutely crap fitter who had somehow managed to struggle and scrape his way to his passing out test job which he was making a right pig's ear of. The problem was that if he failed this particular test piece he would fail the whole Apprenticeship and would be returned to New Zealand, after four hard years, in disgrace, and slung out of their Navy.

My oppo was the deep thinker, I was the good fitter. I'm not bragging, just stating a fact, so what my oppo did was organise the rescue operation of said Kiwi.

First of all he organised a few of us to 'borrow' files and other tools from the workshop and store them in the communal Hobbies Room. The next problem was to purloin the said crap test job before it was ruined completely by our friend. I don't know how it was managed but it suddenly appeared on the Friday evening. The other great problem was that every piece of the test job has to have an identifying stamp mark on it administered with a sharp blow from a hammer on a hardened steel type stamp which left a symbol specifically issued for that particular job on each piece.

If you were going to start work on the face with the mark on it you had to approach the instructor/invigilator to have another face stamped before the removed the original mark. If the Instructor could not find the original mark (you had removed it by mistake), you either failed there and then or were given a new blank piece of material and had to start all over again which inevitably earned you time penalties.

To this day I don't know how it was done, but suddenly, just before the weekend break the relevant steel stamp with the correct emblem appeared in the Hobbies Room, so I spent that night and the rest of the weekend grafting to try and resurrect the mess our buddy had made. I was even brought food and drinks to the Hobbies Room while a guard was stationed outside the door to keep nosy parkers with prying eyes away.

Somehow, at the end of the weekend I'd done my best to rectify the damage, and come Monday morning back at the workshop the test job in question had magically reappeared.

The outcome was that our Kiwi friend had passed his Block Strap Gib and Cotter Test Job with a reasonable margin, to our collective relief. He was so gobsmacked, happy and over the moon that he grabbed me, slung me onto his shoulders and ran from dorm to dorm screaming that he'd passed, to loud applause, the only snag was that him being over six foot tall had me getting almost brained at every doorway. I only managed to avoid severe head injury by lying back horizontally each time.

I was informed at the reunion that the said Kiwi had risen to the dizzy height of Engineering Lieutenant Commander.

Sailing

One of the weekend leisure activities I enjoyed as an apprentice, especially when I was a junior, was sailing.

Caledonia had a veritable fleet of vessels, mostly home made because the Shipwright Apprentices passing our test job was to make a fully rigged clinker-built dinghy so there were loads to choose from. But the sailing fleet also had whalers, cutters and pinnaces.

Weekend activities included rock climbing, hiking, swimming and orienteering, but H and I used to enjoy sailing best of all.

One of the weekend activities now and again was whaler racing on the Forth which was brilliant especially if you were very junior as it got you away from vindictive seniors for a while, and if you were any good as crew the Skipper who was a senior would tend to give you a modicum of protection from them. We had a good Skipper in *Bogy Knight* and we usually won when he was gaffer and I still remember leaning out over the leeward gunnel to try and trim the dish with the windward gunnel taking it in green we were so heeled over.

When there was no racing H and I would sign out a dinghy, take a couple of packed lunches and go up stream on the Forth. This particular weekend we went upstream with another couple of apprentices in a similar dinghy but they decided to beach and have a picnic but we carried on upstream, did a bit of fishing, caught a decent salmon and so rounded of the weekend with a barbecue, when we eventually got back aboard we asked how the other crew had enjoyed their weekend.

Eventually they owed up and explained that they had had a bit of trouble. It appears that where they hauled out was a nice shingle beach and the dinghy was well out of the water so it was safe, so a bit further up the beach they lit a fire with a view to having a brew. The beach itself was cordoned off with a chain link fence and after about half an hour some blokes gathered on the other side of the fence and started shouting and waving so we waved back and thought it was very friendly of them and mashed the tea. The brew was going down a treat when they saw a rather large bloke in some sort of uniform striding purposefully towards them. They were about to offer him a mug of tea, but his greeting was less than friendly. They reckoned his greeting went like, 'Put that f—in' fire out and p— off.' It appears that the other side of the fence was a munitions factory and the workers were more than a bit concerned and perturbed with their fire.

As we became more senior, we still enjoyed our sailing but we eventually needed a bit more of a diversion to fill the weekends. Then H found one in the shape of a girl called Mary he had met while ice skating in Edinburgh.

He was definitely in love and I must admit that she was a very bonny lass and thought when I eventually met her that she looked ever so slightly oriental but I didn't say anything to H. Eventually he got invited up homers and so was I. The only snag was that home was in a place called Gorebridge, a small mining community well to the south of Edinburgh. I don't know how he did it but we got invited for the whole weekend so H formulated a plan.

We recruited a helper (mug), a junior who we promised to look after, sprogging wise, if he would help us, to which he agreed. We got out gear ready Friday evening and set off in a dinghy early Saturday morning and sailed straight across the Forth from Rosyth Dockyard to a secluded little harbour between South Queensferry and HMS *Lochinvar* (Port Edgar). We changed into casual gear and left the mug to take care of the dinghy over the weekend.

We hitchhiked into Edinburgh, pooled over resources and caught a bus to Gorebridge and arrived just before noon found the house and were met by Mary who introduced us to her Stepfather, a retired miner, who was washing his new Vauxhall Victor (rot box). We were taken inside to meet Mary's Mother and I for one got the shock of my life. She was so small, and Chinese, no wonder I thought Mary looked a bit oriental.

Ann, the mother, was a lovely person and was a nurse at the local hospital. We had a great weekend, had a couple or so beers at the local social club on Saturday night and a nice lay in Sunday morning, a leisurely breakfast and Mary's Stepdad drove us back to South Queensferry. We sailed back to Rosyth Dockyard and back aboard. It became a regular trip and we were really well looked after.

Eventually we found out the full story which was quite poignant really. Ann had married Mary's Dad, a Sergeant in a Highland Regiment, and they lived in Singapore. When the Japs overran the place Dad was taken prisoner and Ann, who was now pregnant, was put in an internment camp. Towards the end of the Pacific War Mary's Dad was put aboard a vessel along with a hold full of other prisoners presumably to be shipped back to Japan.

The vessel was sunk by God knows what or who but there were no survivors, just bodies that confirmed the sinking. Come the eventual surrender of Japan and the release of all internees, Ann and baby Mary were repatriated to Scotland and the Dad's regiment organised accommodation and pension and their welfare in Dad's home village.

It became a regular trip to Gorebridge and we were really well looked after, but unfortunately when we eventually joined the Fleet we both lost touch.

It wasn't until much much later that H and I met again and we went back to Gorebridge. Jim the Stepdad had passed away, Mary had married a typical grumpy Jock, she'd had a couple of children and had gone to seed a bit. But Ann was still a lovely little person, the only change in her was that she had shrunk a little bit more.

HMS *LOCH KILLISPORT*

Down South (Dockyard)

I'd enjoyed the final term at *Caledonia* and hadn't meted out the bullying like my predecessors had. I was now out in the grownups' Navy and frightened to death.

I was treated decently as a fully fledged, but green as grass, Killick Tiff. Now I had to put my academic and practical tuition into useful graft and I did it with a will. Initially I was billeted down at Pompey Barracks which wasn't the nicest place to be, but it was good to be able to dump all my gear that I'd humped down from Scotland. It included a full kit bag, a full toolbox weighing a ton, a large pussers green suit case, also full, and a full large holdall.

I met a few blokes I vaguely knew and got to know better over a couple of beers over at the Fleet club, which also made the nights more bearable against the snoring, grunts and groans in the very large dormitory.

Eventually my draft came through, a Frigate, which was undergoing a refit down the Dockyard. One of the blokes I'd met in Barracks was also allocated HMS *Loch Killisport* which neither of us knew a great deal about. So it was with great curiosity and trepidation the next morning when we were bussed from Barracks to the Dockyard main gates and reported to the Dockyard Police Office where, on showing our Draft Chits which they scrutinised thoroughly we were issued with passes, which we were ordered to wear at all times within the Dockyard area. We asked for directions to *Killisport*, and in between sniggers they pointed us down a very wide road and we were told that we would find what we wanted at the third dry-dock on the left.

So off we trotted, full of curiosity, and were stopped dead in our tracks on looking into the first dry-dock. There in all its glory was The Royal Yacht *Britannia*, gleaming like a new pin, the shores (balks of timber) which held her upright were, at the business end, padded with a great thickness of felt to protect the glass like finish paint job from damage. After we'd ogled a bit more, we marched onwards, passed by the second dock which was empty and got the shock of our lives at the third, because down there was the biggest heap of rust you could ever imagine.

It didn't even look like a ship, it had bits hanging off it where Dockyard maties were either slapping on patches of red lead paint, or blowing holes in it with oxy-acetylene torches. With fear and trepidation we went down the brow to an even more horrendous sight and sound, but were stopped in our tracks by a Senior Tiff. He was easily identifiable because he was a lot older than us, wore a very bashed hat and a very badly stained uniform jacket with three very tarnished brass buttons at the bottom of each sleeve, who asked who we were and what did we want. We showed him our draft chits and he instructed us to report to the dock office which he pointed out to us, which was perched on the dock side just forward of the brow. So we clambered back up the brow and went to this, what can only be described as a badly converted garden shed, knocked on the door which was flung open by a Lieutenant Commander who was surrounded by heaps of paper, and barked:

'What do you want?' in a most dischuffed manner, so we just thrust out draft chits at him, at which he grunted, picked up an old fashioned telephone cranked the handle and shouted: 'Get the Chief Tiff up here,' slammed the phone down and shouted at us to wait outside!

Eventually a wizened little fella turned up, his only distinguished feature was a Chiefs hat badge on an old battered hat. He introduced himself as CERA Summers and gave us a Cook's tour of the heap.

The two Boiler Rooms each carried an old type three drum boiler, open front furnace so when steaming the compartments were pressurised by a Fenced Draft Fan and could only be entered and exited via an inter-locked air lock.

The Engine Room was equally quaint, kitted out with a pair of triple expansion, four cylinder open reciprocating main engines, an old fashioned open feed system, two relatively modern turbine driven generators, and a water distillation unit behind the access ladder, but all three compartments were a mess with junk and Dockyard maties scattered all over the place. Eventually a siren blew and the maties just disappeared in one big rush, then me and my new mate, an Ordinance Tiff, joined the rest of the crew wandering down to the main gate, were checked over by the Dockyard Police and were eventually bussed back to Barracks!

Going through the Barracks main gate a huge RPO with the nickname of Tiny shouted at a few of us in a loud and nasty growl: 'You you you and you in here, the rest of you double up away from here!'

We, the chosen ones, ended up as witnessing the weighing off of a poor sod of a sailor who ended up in DQs for being AWOL.

The area where this ritual took place, we were told later, was the courtyard of the cell block that Tiny was in charge of. The area was cobblestoned and the cobbles were polished to a fine sparkle by inmates with tooth brushes while they were awaiting sentencing.

To cut a long story short *Killisport*, believe it or not, was in all respects ready for sea within three weeks, but which time us greenhorns and a few other new larkers were well able to find our way round.

We did a work up day running out of Portland and, after a few incidents, set sail for destinations West via the Suez Canal which was a tale to tell, as we were the first RN vessel to sail the canal North to South after the so called Suez Crisis.

We eventually clued up in the Persian Gulf which was the pits in those days (1968) and had what could be called an 'interesting' period in my Naval career.

Engine Breakdown (Aden)

We were on the way to the Persian Gulf to relieve one of our number of the Armilla Patrol distinguishable by a large black cutout of an Arab dhow stuck either side of the grey funnel.

We, as that patrol, were commissioned to prevent smuggling (fat chance) between the Arab States and Pakistan. En route we had stopped off at Gibraltar which was all very exciting for me as this was my first ship, first trip abroad, first everything, and very keen not to miss a thing. Our next stop was Malta, a very different place to anything I had ever seen before, all dust, sandstone, widows' weeds, and straight streets!

Next stop Port Said, another very different place completely again, all hustle and bustle with everybody trying to outdo the other in all forms, the most prevalent and lucrative being thievery! Us greenhorns had been warned early doors what to expect and what not to do, so we were able to get clear unscathed and into the Suez Canal, which was a real eye opener for me, and an incredible feat of engineering. The landscape is absolutely featureless, just sand, the only break in the scenery was when we entered the Bitter Lakes where the sad sight was abandoned merchant vessels slowly turning dark orange and rotting at anchor. The reason for this was that when Colonel Nasser took over the Canal he closed it and trapped these unfortunate vessels and their crews. With no end in sight the crews were evacuated and the ships left to rot.

Next stop Aden; everything was running so smoothly our Senior Tiff predicted that something bad would happen, and it did. I was watch keeping in the Engine Room as make learn second dicky, when, as we were manoeuvring to come alongside there was a sudden foreign noise which, on investigation, was located and found to be the reversing linkage quadrants of the starboard high pressure cylinder which somehow or other had managed to disassemble itself.

The Chief of the watch immediately rang the Bridge and informed the Skipper of the situation, and advised him to only ring down Stop as the last telegraph order as the linkage would then just collapse. After what seemed like an age and a lot of manoeuvring on the port engine, both main engine telegraphs rang to stop, and then rang down finished with Main Engines, and the whole of the defective quadrant assembly fell into the bilge piecemeal.

So we had an extended stay in Aden where all the engineering staff grafter long and hard to fix it while the rest of the crew got up to all sorts of mischief ashore.

Sea Sickness (General)

Sea Sickness, or *mal de mer*, is indiscriminate. It will strike at the young or old, senior or junior, rich or poor, large or small, male or female, famous or infamous, it doesn't matter who or what you are, if you are afflicted you either live with it and learn to cope, the alternative being not to go to sea.

One of our most revered officers of the Royal Navy, Admiral Lord Nelson, suffered very badly from the affliction so I can class myself as a fellow sufferer. In my case I found out on my first seagoing draft on HMS *Loch Killisport* that I was a sufferer. As an apprentice at HMS *Caledonia* one of my passions was sailing, mainly for pleasure, but sometimes in competition and quite often in bad weather with no trouble at all.

The trip out to the Persian Gulf was another matter entirely, as *Killisport* was not the most stable platform at the best of times. As we left Portland having finished our work-up, destination the Persian Gulf, it was a nice gentle swell with which I could cope, but as we went further west I started to feel a bit queasy, so I went onto the upper deck and saw what was coming I could anticipate the motion and could cope. But down below on watch or day working I used to have to go round with a bucket, as it was very much frowned on to spew up willy nilly.

The worst weather I ever encountered was not the Bay of Biscay although it was pretty bad and my bucket was well used. The worst every roughers I encountered was shortly after departing Gibraltar for Malta with a beam-on sea. As I mentioned earlier Loch class Frigates are not the best sea boats in the world, but with a beam on sea they are almost uncontrollable. My favourite position in those conditions was back against the warm funnel watching the weather coming at you. One second you're watching this huge wave coming straight for you and you look up at it thinking we're not going to make it, then you're going up like in a lift whilst laying over at an alarming angle, and the next second you're sliding down the other side into a very deep trough again at a crazy angle, but at least I could predict the movement and brace myself accordingly.

Then it was my turn on watch down below and then I knew I was in real trouble, one hand for me, one hand for my bucket, and wish to God I could curl up and die!!

Eventually I managed to cope a lot better and just suffered nausea and headache. But when I joined *Dreadnought* we dived when we reached the continental shelf and even if you were only at periscope depth it was nice and calm. The downside to this was that when you returned to base the boat surfaced on reaching the continental shelf, which doubled my trouble!

Suez

Hanging about in Pompey Barracks was no fun. I was in a strange place, knew nobody and was very apprehensive. I'd left friendly *Caledonia* fully booted and spurred complete with hammock, tool box, green canvas suit case, and Pussers hold all.

I was a lowly fourth class Tiff, a Killick, the lowest of the low again, and had been given a draft chit to HMS *Loch Killisport*, but I wasn't due to join her for another week. So I did a bit of research and found out a bit about my new home to be.

It was a Loch class anti-submarine Frigate. Armament was a four-inch gun forrard with hedgehog a forward firing multiple depth bomb thrower. Back aft were four two pondered on sponsons, and it was built in 1944–5.

Killisport, I found out, was in dry dock undergoing a refit to get ready for a stint out in the Persian Gulf. Eventually I was allowed through the Dockyard Gates and was accompanied by a third class Tiff who also drafted to *Killisport*. We strolled past the first Drydock and there in all her glory was the The Royal Yacht *Britannia*, absolutely gleaming in the early morning sun, a radiant jewel.

We moved on a couple of drydocks and came across a sad looking heap of rust and dirty grey paint with the pennant number just discernible, F 628. This unlikely looking load of junk was *Killisport*. We went aboard and things didn't get any better, with air hoses, gas welding hoses and bottles, arc welding cables. Come stand easy we assembled in the Tiffs Mess which was quite tidy really, and we met a few more Tiffy crew members and we were invited to bring our tool boxes down and get stuck in to help out. Which is exactly what we did, and I quite enjoyed myself and managed to find my way round as well.

I couldn't believe it, but we were up and away in a fortnight and doing sea trials out of Portland and I was learning how to be a proper Tiff. The only problem I had was that I found out that I was very prone to seasickness and my Harbour Station was the tiller flat, battened down with a leaking reciprocating steam engine steering gear. Leaking steam and the smell of warm damp tarred rope, that smell still turns my stomach.

Our work up complete we stored ship back at Pompey. Took aboard a detachment of Royal Marines and away we went. We called in at Gibraltar for a couple of days and then on to Malta for another short stay and then the Suez Canal. So far I had been amazed and overawed with everything which was all new to me, but the next stage was going to be at least, very interesting as this was the first time that an RN Vessel had gone from North to South through the Suez Canal after the Suez Crisis when nasty Mr Nasser caused trouble.

Royal Marines featured significantly in the conflict and our detachment had lost a few good mates, so there was no love lost. As we secured alongside, ready to transit the canal in convoy we were assailed by locals flogging all sorts of junk and toys. And the brow was mobbed by locals purporting to be the official Pilot so that they could get aboard to thieve anything that wasn't nailed down. But the Skipper had anticipated this and had arranged for the Marine Detachment to wear full dress uniform and man the Quarterdeck and the bow. The uniforms alone deterred a lot of the locals, but some of the more persistent were left in no doubt about the consequences of misbehaving.

On the Quarterdeck, which in this type of vessel had quite low freeboard, the local touts could put their wares mostly clockwork trash on the deck but if they tried to retrieve them too early a highly polished beetle crusher would descend rather heavily on searching digits. And so to the canal where we flew the biggest Union Jack at the bow, and similarly on the stern a huge White Ensign to be sure that the natives had no doubts as to the vessel's nationality.

To reinforce the message even more forcefully the Bootnecks manned the Quarterdeck sides in full dress uniform, and as soon as the native rubberneckers saw them they scooted as if they were on fire.

And so to the Persian Gulf, but that's another story.

Crater City

I was a lowly 4th class Tiff aboard HMS *Killisport* enroute to the Persian Gulf.

We had to call in at Aden for repairs as the HP Cylinder Stephensons Link Gear of the stab'd main engine had collapsed in a big heap, and required major parts remade and fitted.

It was in the days of Mad Mitch and the fight against terrorism in the area around Crater City, and a run ashore was out of the question as it was off limits, so we ganged together and went to see what all the fuss was about.

The area generally was dusty and dismal and reeked of poverty and not at all what we expected. To cap it all beggars were everywhere, old and young. The one bit of advice we got from the old hands was not to give to beggars under any circumstances, but the further we went the state of the beggars got worse and worse. One young lad we passed was as thin as a rail, his legs were in what looked like a reef knot dressed in rags and he shuffled along on his knuckles. His skeletal face pleading with passersby for anything of value. I couldn't resist and passed him a few Annas that I had in change as secretly as I possibly could.

The next thing I know is that he's on his hind legs and absolutely galloping away. The second thing that happened was that the rest of his oppos completely mobbed us and we had to leg it back aboard, and I was in the tish for ruining a perfectly good run ashore.

Gulf (Chief Tiff)

In the previous chapter it was explained that the Persian Gulf was the pits in those days. There wasn't even a decent berth alongside at Bahrain which was our 'Base'. We had to anchor off and use liberty boats if any of us wanted to go ashore to see nothing much. The only ones who habitually did so were the Rum Rats and Winos because there was a sleazy looking boozer just off the jetty which was owned and run by a Southern Irishman who preached to anyone who would listen about the Treacherous Northern Irish and the Dreaded English. But typically Irish, he would serve anybody if they had the readies!

Anyway our duties were to patrol the Gulf from top to bottom which wasn't too bad at times because we sometimes called some of the nicer places. One of those was Kuwait, where the resident Brits always offered our crew up homers whenever we were in the area. It was always individuals, or couples at a time, and even provided transport to and from their Residence, I think that deep down they craved a bit of news and gossip from home.

Me and my mates' trip on our grippo run ashore, we found out later on while swapping stories, was typical. We got tarted up in the afternoon and in the evening requested of the Duty Officer permission to go ashore and after a quick inspection, to check smartness, it was granted. So off we trotted to the foot of the bow, turned aft, chopped a salute, turned right and zoomed down the brow to be met by a smartly dressed Arab, who pointed us towards a smart, but not over-large, limousine and offered us the rear seating which we readily accepted. Our driver then decided to offer up a Moslem prayer (eyes down for line and house) then calmly got in, started the engine and WAM, he drove like lunatic, flat out all the time, corners, hair pins, stop signs, it made no difference, he was a loony driver. It was later discovered that the ritual on the jetty was a prayer to Allah which would absolve him of all blame should an accident occur as it was Allah's will.

So the commission went on, showing the flag, trying to stop smugglers trafficking in either arms, treasures or people, to and from the Persian Gulf States and Pakistan. If we tried to stop them they just put up two fingers to us, opened the throttles on their twin supercharged monster diesel engines and away they creamed, giving us not a cat in hell's chance, even with the standby boiler up to a full head of steam and both engines at full chat.

Therefore we ended up just tootling about, showing the flag and generally being bored. The food we had wasn't bad, the Ceylonese chefs did their best and augmented our diet with line caught fresh fish and any other fresh produce they bartered for from the natives.

The very greatest highlight for the whole crew, living this boring life, was Tot Time, an eighth of a pint of best pussers rum, it gave you a rosy glow and made you able to eat almost anything with gusto! That being so, the Tiffs' Mess members were not happy bunnies as for quite some time us juniors were coming up short. The Chief Tiff (and Chief of the Mess) at first dismissed the shortfall as a fluke or genuine error which us juniors, who were the ones that were coming up short, didn't agree.

The rest of the Tiffs in the Mess tended to agree and a couple of the more senior members questioned our Mess man, a junior seaman, assured them that he always made sure that the measure was right. So these Seniors went with the Mess Man to oversee a few days of issue with permission of the Issuing Officer, and reported back to the members that all was shipshape and Bristol fashion, but we still came up short.

It wasn't till we had a bad engine breakdown, and we were all hands to the pumps as they say and carried on past Tot Time to finish. The Chief Tiff said he'd been called away and would rejoin us as soon as possible. Our Senior Tiff had to leave us to it for a little while to get some dimensions he had written in his pocket book that he'd left in his jacket pocket. On approaching the Mess he noticed that the door was about a quarter open which was unusual, so he approached cautiously thinking it might be a tealeaf. What he saw was our 'beloved' Chief Tiff helping himself to his tot plus another one, both full to the brim.

Our tot measure was not the usual type where when it was filled to the brim it was a full tot, a design that was fraught with disaster when decanting to the glass because if your hand was unsteady or you rushed it you got your own back and wasted some of the precious liquid. Our design, furnished by the Chief Tiff, was what you might call a bifurcated shape, the bottom section volume being a full tot, tapering in from the base up. A second reverse tapered section a bit more than half the volume of the first was silver soldered to the bottom section. As our Chief explained, when decanting your tot from the supply jug, a bigger target area presented itself for filling, and if overfilled by accident the excess could be decanted back to the jug.

Our Senior Tiff confided in his more senior Mess Mates explaining what he had witnessed and had devised a plan which his mated approved of so he set to work and his plan was ready for the next days Tot Time. As usual 'Chiefy' took his tot first, but in the past we hadn't noticed anything irregular or untoward but he always insisted that the tot decanting station was in the far corner of the mess. Therefore as you measured out your tot nobody could see exactly what you were doing, but nobody had any thought of cheating.

Came the day of days, the senior Tiff made sure all mess members were present. The tot measure was always sat on a large deep saucer in case of accidental spillage so that it could be returned to the jug with no waste. Our beloved Chief Tiff, up first as usual, soon became a bit agitated, then frustrated and then angry, but by this time the 'G' was not only the right level in the measure, but the saucer was overflowing so much that it had spilled onto the table and was dripping onto the deck!

Every mess member gave a hearty cheer and Chiefy exited very quickly, very red faced. But all was not lost, our senior Tiff had squared it with the Jimmy and the Coxswain and the ration was re-issued. The Chief Tiff mended his ways drinkwise but was not much liked anymore, and on the grounds of ill health once back in the UK.

The secret of the senior Tiff's success was that at the join of the measure sections he had drilled lots of minute holes so that if you overfilled even slightly the saucer caught the spillage.

The Still

Back in the Persian Gulf with HMS *Killisport* one of the main bones of contention was fresh water!

The shore supply from our base in Bahrain was inconsistent and always very brackish, so much so that it was undrinkable and only used for some tasks in the Galley like supplying the spud peeler, we even showered in sea water because the onboard distillation unit could not cope with the demand.

We relied on this unit for drinking water and more delicate jobs in the galley and the sick bay, and of course boiler feed water. This distillation unit was situated in the Engine Room in the centres behind the deck plates access ladder. Its function was to take in sea water, boil it up and condense the steam formed to produce rather bland but pure water. The resultant brine formed in the process returned to the sea.

It was a continuous process, apart from having to descale the internal coils quite frequently and involved one boiler being kept on line to supply steam to the boiling coils of the still, so the Boiler Room had to be fully manned 24 hours a day regardless of deployment. Also a continuous watch had to be kept on the distillation unit so another part of the engineering department was involved 24 hours while the other departments took things easy.

At the time of the incident we were anchored off Bahrain again and running auxiliary as usual to supply decent fresh water which involved a Chief or PO in charge of the Boiler Room and a leading stoker to change and clean fuel oil sprayers as required, and also a Killick Stoker in the Engine Room looking after the still.

Everything was quiet in the Tiffs' Mess as it was just after tot time when there was a gentle tapping on the mess door. Our mess man answered the door and relayed the message that the still operator would like a word with the duty Tiff.

'Asked him what he wants?'

'He says the still's fallen over.'

'Well tell him to piss off and fix it.'

The mess man passed on the message and shut the door.

I must at this time point out that in Naval parlance 'Fallen Over' meant broken down!

About fifteen minutes later there was another knock on the mess door, a little more urgent this time. The mess man answered the door again and it was the still watch keeper again asking for the Chief Tiff this time. He was dozing off at the time and told the mess man to tell the guy to piss off and fix it or he'd be in the rattle and again the mess man passed on the message and slammed the door in the operator's face.

Ten minutes later there was a hammering on the mess door and it burst open and a red faced Killick Stoker marched up to the Chief Tiff and said: 'Chief the still has fallen over and is now leaning at an angle of approximately thirty degrees to starboard and I've had to shut it down because the main steam supply line has fractured!!'

With the revelation we all took notice and went down to the Engine Room to survey the damage and sure enough the main support girders had rotted away on the starboard side and the whole thing was in imminent danger of slipping into the bilge.

It became a major repair job for all the engineering department and until it was fixed we were on shore supply brackish water which wasn't even suitable for boiler feed.

Needless to say the still watch keeper was showered with apologies, and a few tots.

Muscat (Funeral)

We the ERAs. of HMS *Killisport* eventually got the defective main engine up and running and were soon on our way to the Persian Gulf once more. We'd got as far as the narrow entrance to the Gulf when suddenly we did an about turn and started back the way we'd come.

Eventually the Skipper got on the Tannoy and announced that an RN vessel was required at Muscat to perform an official duty to be explained later! We came alongside a couple of days later and shore leave was granted for those off watch, so we donned our tropical rig of white hat cover, white shirt and shorts and white Blancoed canvas shoes, and we all looked rather tiddly and smart.

We were met by a rather dour looking detachment of Royal Marines who were based in Muscat, and one of our number asked if by local customs there were any taboos, to which one of their number replied that bare knees were a No-No!

Eventually one of us was brave enough to enquire if any of them knew why we were here to which one of their number explained that a burial at sea was required for one of their number. No wonder they were a bit upset and glum.

It appears that the poor unfortunate was on point while carrying out a patrol in the desert with a squad of his mates; they were ambushed and he copped it first. The attack was beaten off but their mate was very badly wounded, and although they got him back to base he eventually succumbed to his wounds.

Unfortunately there were no refrigeration facilities available in that area therefore time was of the essence. So their unit of the deceased's colleagues, combined with our own detachment, came aboard, all in full dress uniform, a very impressive turnout, complete with the Royal Marine Padre who was to conduct the service. The appropriate position was reached and we came to a stop on a glass calm Indian Ocean with a gentle swell. The whole of the Service was filmed for the benefit of the deceased's family and friends. Last Post was sounded, the Padre gave a fine eulogy and finished with a prayer, and the body slid into the Ocean with barely a ripple, then the Bugler sounded Reveille, and it was all over.

We returned to Muscat to disembark our passengers and went on our way, but you could not get the odour of death out of your nostrils. We tried all sorts but the only thing that tended to help was, believe it or not, cockroach killer spray which had a distinctly orange peel zest odour, and to this day when anyone peels an orange it reminds me of that sad miserable episode.

Jellyfish

I was a lowly fourth class Tiff (PO) on the Armilla patrol (HMS *Killisport*) and we cruised the Persian Gulf, attempting, badly, to prevent smuggling and slavery. Badly because as soon as we tried to stop a real transgressor they flashed up their high power engines on old looking dhows and creamed away at high speed beating our measly 19 knots flat out that took us about half an hour to achieve.

The Oggin in the Gulf was shallow and warm and contained a myriad of creatures of exceptional size. One of our Dabtoes was an archery expert and used to go fishing with a bow and arrow, the arrow tethered to a length of fishing line, and I have seen him impale two edible fish on one arrow.

Our Ceylonese (now Sri Lankan) Chefs used to supplement our rather boring diet by catching very large mackerel and preparing and cooking them by cutting them into one-inch thick steaks and frying in butter. The method was that the first side was only half cooked while the top side was sprinkled with a mild curry powder, the steak was then turned over and the part cooked side sprinkled with curry powder and turned back over to complete the cooking process. Absolutely delicious!

But I digress, the incident that is still fresh in my mind is the time we steamed into a large shoal, collection, mass, or shower, whatever you call a large bunch of jellyfish.

Killisport was an open reciprocating triple expansion four cylinder twin engine Frigate with Stevensons link reversing gear of Second World War vintage. The main and auxiliary bearings and reversing sheaves were lubricated with soluble oil. The sheave pan were lubricated with soogie, a soluble oil and water mixture, but the sheave pans tended to leak a bit and the lubricant ended up in the bilge where it tended to separate out.

But I digress, we hit this mass of jellyfish and we ended up with vacuum trouble and warning alarms going off, first one side and then the other. To explain, exhaust steam from the main engines was recycled by cooling it in the main condensers and converting it to water and returning it to the feed system to be turned back to steam in the main boilers.

The main sea water circulating pumps passed cooling water via large plate filters through the condensers and back to sea.

There in lies the problem, the plate filters kept blinding with jelly-fish and consequently lost vacuum and therefore efficiency. The routine adopted was to stop and isolate the main circulating pump, withdraw the filter plate and remove the jellyfish. The Chief Tiff said that they would melt so with blind faith we chucked them into the bilge. As soon as we cleared one side and got it back on line the other side packed in. It took forever to clear the infected area.

To compound the situation another problem that had not yet been uncovered was that the galley spud peeler discharge went overboard via the Engine Room bulkhead and an inboard bend had eroded and corroded through and discharged the majority of the spud peeler mush into the Engine Room bilge.

Shortly after clearing the jellyfish horde we returned to anchor off Bahrain after a flat clam transit. After a few lazy days we set sail for Kuwait and the sea got a bit choppy and suddenly the pen and ink in the Engine Room was absolutely terrible. The cocktail of salt water, soluble oil, rotting jellyfish and rotted spud mush permeated the whole of the vessel and made some crew members violently ill and turned number one uniform gold braid and buttons purple.

The Disappearing Ky

Ky was a ritual for the silent hours, the middle and morning watches. It was almost as sacred as the Tot.

Ky was the pussers version of cocoa or hot chocolate and was made by painstakingly finely flaking a large block of dark unsweetened chocolate into a large fanny, then large quantities of sugar and condensed milk are added and topped off to final quantity with water (not a lot) then the whole lot is boiled up by injecting live steam from a drain in the Engine Room which is kept especially clean for the purpose. The drain we used on the old Loch class Frigate was the steam drain off the operating cylinder of the port vacuum pump.

Came the middle watch and Ky time the Engine Room watch had just finished theirs when a lot of manoeuvring orders started coming down and the watch keepers were all a bit preoccupied when a young sailor came down with the upper deck watches Ky for boiling up. He asked one of the lads where the steam drain was and a vague finger pointed in the general direction, not realising that the lad was new on the job. The next thing that happened was that the port salinity alarms went off, indicating contaminated boiler feed water which was very serious as the boiler could prime and cause real bad damage. Then it just so happened that the young sailor came back into the centres looking confused and tearful.

'What the hell's up with you?' the Chief of the watch barked.

'Me Ky's disappeared!'

'Oh shit, dump the port vacuum pump discharge fast,' he growled, and the killick stoker of the watch leapt to open the drain valve.

'You stupid git you used the wrong line, you got the vacuum side, piss off out of my sight.'

'What do I do about the Ky?'

'Tell your watch Chief what happened and expect a good kickin',' and he carried on trying to clear the salinity alarms which had been set off by the contaminating Ky.

Basrah

As I've explained before, *Killisport* was THE Armilla patrol and it was at the time when Iran and Iraq were sabre rattling, which was well before the real bust up and real war. We were anchored at the mouth of the Shatt el-Arab at the north end of the Persian Gulf, which is the confluence of the Tigris and Euphrates. Anyway it was as hot as hell and we were at a half hour's notice for steam so we had to turn over the main engines every ten minutes, which meant that the Boiler Room had to be fully manned all the time as well, and the Dabtoes were sunning themselves and fishing on the upper deck, with us down below sweating our bits off. We were told it was necessary, as if the shit hit the fan we were to immediately steam up river and evacuate the British Nationals.

Anyway it all calmed down eventually and as a thank you for our vigilance we were invited to Basrah and me and a couple of mates got invited up homers to this beautiful villa which boasted a swimming pool and air conditioning, the works, heaven.

One of the snags was that the man of the house, a high up oil company executive, was away in the UK on business. The second snag was that the lady of the house, although clocking on a bit, was very frustrated and man daft if you get my drift. The third problem was that she was an avid fan of Tommy Steele and had all his records and was a founder member of his fan club. The fourth and main problem was that she thought that I looked the spitting image of him and wouldn't leave me alone, especially in the swimming pool, and I had a terrible time fighting her off, much to the amusement of my so called oppos, and was greatly relieved to escape from the house. Her parting gesture was to invite me back, but on my own next time!!

Diving

Killisport was anchored off Bahrain, as in those days there was no deep water harbour, the run ashore was dismal with nothing to do and nowhere to go, so most of us stayed aboard consuming loads of liquid and chucked the empties over the side. These bottles and cans were of some value to the impoverished local community, and pearl divers used to come out in dugouts and small dhows to dive for them around the anchorage using a very ingenious method. A pole was rigged over the side projecting out about a couple of feet. A large rock was attached to a length of rope. From the crew of two the elected diver jumps over the side and his mate hands him the rock and the rope is looped over the pole, the diver grasps the bight of the rope and stands on the rock. When ready the diver takes a deep breath, lets go of the bight and allows the rock to carry him to the sea bed where he steps off and swims around gathering anything of value before he has to surface for air. In the meantime his oppo has hauled up the rock ready for the next dip.

One amusing incident concerning these activities involved our own vessel's Shallow Water Diver. Every RN vessel carries at least one qualified Shallow Water Diver for obvious security reasons, and they have to keep qualified by doing regular diving training exercises. Out in the Persian Gulf the water is very often blood heat but it is recommended that a suit be worn to protect the wearer from jellyfish and other stinging creatures. One particularly painful animal looks very much like pulled out orange knitting which tends to drape over the body and stings everywhere it touches.

Anyway this particular time the diver dressed in a black drysuit and re-breathing set, told his oppos to keep an eye on a couple diving off the starb'd side and he went over the port side. He dropped to the bottom and hugged it, approached the pearl diver from the rear and tapped him gently on the shoulder. Relating the incident later, our diver said his target's head swung round, caught sight of this big black thing, his eyes nearly popped out like a cartoon character and he shot to the surface as if jet propelled. According to the rubbernecks on the quarterdeck, the guy shot straight out of the oggin into the boat as if he was on fire and paddled like hell for the shore.

Sunburn

H told me about the time he ended up in jail in Cape Town. It appears that a group of Chiefs and POs had gone on a run ashore together and had clewed up in a night club and H had pulled this lovely tall slim sunburnt lass and eventually he was invited up Homers. So off they went with H thinking his luck was definitely in, strolling along with their arms around each other. Then suddenly a police siren sounded and then a cop car screeched to a halt in front of them. Two Yarpy Scuffers jumped out, grabbed him and bundled him into the car and shot off to the Police Station and bunged him in a holding cell with a crowd of ne'er-do-wells. He was allowed a phone call so he called Bigbury Bay and got hold of his DO who promised to try and sort something out. H— explained that he hoped it wouldn't be long because he was banged up with some very unsavoury characters.

Eventually his DO turned up with an official from the British Embassy who asked him what had happened so he told him as it was, and the guy went off to have a word with the fuzz and came back with the DO grinning like Cheshire cats saying he was free to go. So what was it all about? he enquired. Well, the Embassy wallah explained that he was in civvies in the RN and was not aware of the law. 'What law?' he enquired. 'The law regarding segregation and apartheid, you're not allowed to consort with coloureds.'

'She wasn't coloured,' he protested, and it was explained to him that she was classed as a Cape Coloured, a half breed, and he was shopped by one of the bar staff in the night club where they had met.

'Well it's a stupid law 'cos she was gorgeous and I thought that she was just well sunburnt, what a bummer.'

Collision

Whilst serving on an old Loch class Frigate out the Persian Gulf we did a show the flag trip to Muscat which was a bit of an eye opener, but it was either fortunate or unfortunate depending on your point of view. We got an urgent signal to head out into the Indian Ocean to offer assistance to two huge tankers that had collided.

The two tankers involved were *Ferdinand Gilabert* and *Melika*, two modern tankers that had collided in a huge ocean in broad daylight. The *Melika* had a hole in her port side that you could park four double decker busses alongside each other. She'd caught fire back aft so the crew had abandoned here and left her steaming into the distance. The only reason she stopped was that the oil fired boiler ran out of fuel. *Ferdinand Gilabert's* bow had been completely stove in and only the deck was still intact, but flapping badly in the swell.

I got a double wammy from the incident because I as a junior Tiff, was detailed off to accompany a senior Tiff to repair the fire damaged fridge gear on the *Melika* so that it could be re-stored because the crew wouldn't return aboard until it had been done. The job was a pain in the bum because apart from having to climb aboard up a never ending rope ladder, the fridge pipe work was lagged with cork and we both ended up as black as the hobs of hell, good stuff burnt cork, ask Al Jolson.

Then once that was done we were ordered to tow *Ferdinand Gilabert* to Karachi. Because of the damaged bow we had to tow her stern first to prevent excess pressure on the now exposed bulkhead. The whole crew turned too to hump anchor cable back aft, two links to a man, to weigh down the tow line. The snag was that her rudder was jammed hard over to starb'd, she kept sheering over to our port to an alarming angle before shooting back on track and then doing it all over again. To try to keep on track and combat this effect we had to run the port engine at maximum revs and starb'd just ticking over. Then I got the second wammy because they wanted me to be part of the boarding party to try and centre the rudder.

By this time we'd been joined by a couple of Destroyers and a Carrier and they were going to shift us by helicopter and land us on the floppy foredeck. The helicopter winched our gear and tools aboard and then the plan was changed and we all lost our gear. Another set of lads were landed on and eventually centred the rudder, but found that the fresh water was contaminated so they broke into the bonded stores and quenched their thirst on wine. It appears that by the time we reached Karachi they were all ratarsed. When we eventually got to Karachi the Port Authority refused us entry because our charge was leaking a small amount of fuel, so our Skipper informed them that after a very difficult towing operation, if they did not accept the vessel he would put a four inch brick in it and sink the damned thing in the harbour entrance shallows, so the authorities allowed us to leave it with them.

The good bit that came out of it was that I got £35 as a 4th class Tiff which was a lot of money in those days. I'm led to believe that the salvage money for this occurrence is still the highest paid to the RN and was in the multi-millions, and as a bonus we were all presented with pewter tankards because we refused cash for donating blood to treat the injured in the accident.

Karachi (Docking)

I've never to this day found out why Their Lordships in their infinite wisdom wanted *Killisport* to undergo a dry docking and major overhaul in Karachi. But that is what they wanted so we had to do it, much to our dismay and apprehension, because the Pakistanis are keen as mustard to learn, but a little too OTT at that time (we are talking late 50s).

Came the appointed time, we steamed from our patrol area in the Gulf to Karachi. We shut everything down, and were docked down quite smoothly, efficiently and quickly.

We were told that we would still be living aboard as all the normal services, fresh water, electricity, and sewage disposal, would all be organised ashore, and that's where it all started to go awry!

Our Engineer Officer decided that brass and the bronze valve hand-wheels were too much of a temptation for the local tealeaves, so he had them removed. Unfortunately he didn't insist on them being labelled, so they were just securely stored in a bid heap which turned into a nightmare when it came to refitting the right hand wheel to the right valve. This problem was self-inflicted.

The non-self-inflicted problems then came thick and fast. Firstly the sewage line to shore had to be blown clear daily. Unfortunately the automatic non-return valve didn't work and we got our own back, all over the place. Fortunately for us it was the Dockyard's fault so they had to do the clean ups!

The shore power supply was very intermittent, with us suffering frequent blackouts. Now in a normal domestic situation, if there is a power failure there's a 50–50 chance that it would be daylight, but if it's night time there's the street lights, the moon or emergency candles. But in a ship, down below there is nothing, and unless you permanently carry a torch you are completely helpless because the darkness is absolute, and can be very intimidating.

The final straw was the fresh water shore supply which was not only brackish, but also very intermittent which wasn't too inconvenient for the majority of the crew, but for our Chefs it was a nightmare, and regular flare ups between our Cantonese Chefs and the Pakistani water supply 'Engineers' tended to be quite spectacular, with both sides talking and nobody listening.

All in all this trial docking was deemed not a great success, and Their Lordships decided against any repeat exercises of this nature, thank the Lord!!

Cockroaches (Pests & Pets)

Cockroaches – you either tolerate them, or hate them, but in the Navy, in my time, LC were a fact of life, and they are survivors! On *Killisport* we had first and second commission cockroaches, you could easily tell them apart because the first commission ones were spray painted.

Their Lordships brought out an anti-cockroach aerosol spray can which had the odour of orange peel which was quite pleasant really, but the cockies hated it and was deadly to them.

Killisport being a bit ancient, the accommodation compartments had fulfilled many roles and some of the adjoining compartment bulkheads had various holes drilled to mount bits and pieces on both sides from time to time. In fact some of the bulkheads were like colanders they had that many holes in them, and what the occupants used to do before delousing their compartment of cockroaches was to find out where the holes were that had been blanked off with sticky tape on the other side before painting. The idea was to find the blanks and poke them out with a pencil or similar because when you did the spraying of the orange zest stuff and sealed your door shut the roaches used to escape through the holes you'd made in next door's bulkhead and after sufficient time had elapsed to ensure all your resident roaches had escaped to next door you dashed in, sticky tape at the ready and resealed all the holes you had previously made and gave your new patch a coat of matching paint to cover up your delousing efforts.

Sometimes, when we were a bit bored, uneventfully tooling about, some lunatic would organise a Cockroach Derby, and organise it like a real horse race, either on the flat or over the jumps with some wide boy running a book and fixing the odds. The snag with cockroaches is that they appear to have very little sense of direction, and go in fits and starts, so the race course has to be constructed in very defined lanes, and if steeple chases the obstacles have to be built as well. Also the little devils are quite agile and sometimes attempt to change lanes, so the owner has to be very vigilant or end up being disqualified.

Identification is by a dab of paint on the back of each entrant the colour to be chosen by the 'Owner'. The Owners have their own ways of training, some starve their entrant and then encourage it to chase a stick with a morsel of his charge's supposed favourite food on the end. Others are a bit more brutal and just encourage them with a quick poke with a pointed stick now and again. Quite heated arguments can ensue, but most disputes are resolved amicably.

One of the more bizarre things to do with cockroaches was to keep them as pets! While out in the Persian Gulf most of the consumable stores were shipped out to us from the UK, but some items such as fresh veg were grown locally in places like Kuwait, and very good the produce was too. Unfortunately (or in some bizarre cases fortunate), some of the local greens, especially cabbage, attracted a variety of cockroach that was a monster of the roach world, measuring between two to three inches long! Some of the lads tried to train them with varying degrees of success. But some wag, we never were very sure who it was, found this absolute monster and managed to construct a small mesh cage for it, and jammed the cage on top of the cable tray work, and situated it above the communal water cooler fountain outside the Canteen on the Canteen Flat, a very popular area.

All the crew eventually knew about it, and some used to feed it, others used to talk to it, and typically it would eat anything and everything and just kept on growing. We never did find out its final fate, but it had a very 'fulfilling' life while it lasted.

One sure-fire way of collecting and getting rid of cockroaches was to trap them. What you need is a jam jar with an internal horizontal surface below the neck. Put some instant coffee granules into the bottom of the jar and lubricate the inner horizontal surface with mineral oil or light grease and leave it in a known cocky thoroughfare. Cockroaches just love coffee and dive in but can't get out because of the lubricated upside down surface.

The best way to get rid of them is to chuck the whole lot, jar included, overboard, or as in RN parlance, give it the float test, but make sure your vessel is underway. The reason for this is that we were at anchor off Bahrain again and the water surface was like a sheet of glass. Our mess man had set a coffee cocky trap a while ago and was now well full, so he decided it was ready for the float test, so he went onto the Quarterdeck and gave the full jar a good over arm throw, and stayed leaning on the guardrail waiting for it to sink and as it did a sort of black stain appeared on the water surface which fascinated him.

Something else distracted him for a while and when he looked back the Black Blob seemed to be a bit nearer so he looked more intently and lo and behold it was getting closer. The cockroaches were swimming towards the ship and he was fascinated and called a couple of his oppos across to see what would happen.

The dammed things got to the ship's side and climbed up to a scupper and disappeared, once more to inhabit HMS *Loch Killisport*.

The moral of this story is – be sure to be underway when disposing of unwanted guests.

Hammocks (Danger)

My first commission was an old Loch class Frigate HMS *Killisport*, and I think we must have been one of the last vessels in the RN to sling hammocks. Slinging a hammock was a pain in the bum, but once rigged properly it could be quite comfortable, but fraught with disaster if not. Poor rigging could cause all sorts of pitfalls, for instance if the nettles aren't tied to the canvas eyelets correctly they slowly let go one at a time while you are crashed out. Eventually the last couple give way and down you go. Which end gives way first determines which end hits the steel deck first. If it's the head end a severe headache is guaranteed. Another problem is that even if the nettles are correctly tied but left too long or if the stretcher is too short, as the occupant tosses and turns eventually the stretcher lets go like an arrow from a bow and is lethal for the neighbours, and also buries the occupant as the hammock sides close in on him, which makes it difficult to evacuate. Some who thought to make their lives easier by using slip knots to speed up the rigging and unrigging of their sleeping appliance often came unstuck by their own messmates who, coming aboard from a good run ashore, a little merry and playful, would ease the slip knots undone, place the loose ends in the slumbering occupant's hands and then wake him up, and stand back to watch the poor unfortunate seesawing until his grip gave way with bone-crushing results.

An alarming incident occurred on *Killisport* while transiting the Indian Ocean one early morning just before reveille when one of the Ceylonese chefs, for some obscure reason, went berserk with a very sharp carving knife in the Chief and POs' mess. He ran amuck underneath the rows of hammocks slashing the bulging undersides crosswise, fortunately not drawing any blood, but it was quite a sight with so many arses poking through.

The worst bit really, for us non-sailor types, was the five minutes seamanship last thing at night unlashing and rigging the damned thing, and first thing in the morning lashing and stowing it. You were supposed to have seven lashings round it to ensure a neat and tidy tube effect with all your bedding and cordage inside, and neatly stowed vertically in the hammock netting. Usually our Tiff's mess netting looked like a scran bag full of sacks of spuds with bits hanging out.

Fred

I'd just returned from the Persian Gulf and was back in Pompey Barracks and immediately came across an old apprentice classmate. We decided that a run ashore was in order, had a few wets and a good natter. He'd been out to South Africa on another old Bay class Frigate, the same vintage and build to my ex-Loch class. He'd been in trouble down there due to their apartide laws, but that's another story. The conversation then turned to my Persian Gulf time.

'A towney of mine, a Bootneck, was out there about the same time as you and he reckoned that some crazy sailor and him built a canoe out of an old packing crate.'

'That crazy sailor was me, and you're talking about Jumper Collins.'

'Bloody hell you're right. It's a small world ain't it, did you really build a canoe?'

'Yes, it's all a bit bizarre really, we were anchored off Bahrain which was a rubbish run ashore in those days, but there was a Canteen just off the jetty. This particular evening our Chief Stoker went ashore and got absolutely ratarsed and on his way back aboard through the yard saw a self-contained air conditioning unit, and being a big lad he thought, in his confused state, that the Captain might like it in his day cabin. So he hoisted it onto his shoulder and carried it on to the end of the jetty for the last Liberty Boat which was waiting for him. He peered down, saw the boat, shouted below and dropped it aboard from a great height. The reason for it being from a great height was the fact that the tide had gone out and the AC unit hit the prop shaft at a rate of knots and bent it like a banana. Another sea boat had to be sent to tow the damaged Liberty Boat back, together with the Chief Stoker. Chiefy was on Captain's report on a double wammy of being adrift and damaging pussers' property. What did not help his cause very much was meeting the Captain first thing the next morning when the Skipper was having his morning constitutional and saying to him in a jovial fashion.

'Good morning Sir, you look about as bad as I feel.' Which went down like a lead fart, and for which he was suitably punished, but we all had a good laugh about it in the end.

The sequel to the incident was that a new prop shaft had to be requisitioned and was eventually shipped out to us, and caught up with us when we were tied up alongside the jetty in Kuwait.

I was the Tiff who was detailed off to fit the new shaft, and possession being nine tenths of the law, I commandeered the packing crate it came in. I took it apart very carefully because, although it was only softwood, it was good quality and reasonably knot free, and just the right length for a two man canoe. I conned a Bootneck (Jumper Collins) into helping me, but I think he was keener than me. We borrowed a rip saw from the Chippy and laboriously and sweetly cut the planks into strips to make stringers, and bummed marine ply offcuts from the Chippy for formers (frames) and canvas and paint from the Buffer for skinning and waterproofing, and we proceeded to build a canoe.

Most of the crew thought we were crackers, some took the urine, and some were amused and kept asking what we were going to call it. Being thoroughly teed off with stupid questions, in desperation one day when Jack Dusty asked Jumper the same question, without thinking he blurted out 'FRED'.

'Fred?' Dusty almost shouted, and we all thought he was making it up.

Yup, Fred it was, and from then on the name stuck and rubberneckers would regularly come by to see how Fred was coming along.

We eventually gave it a good coat of paint and Jumper did a fair job of painting Fred Flintstone on the bows, a couple of paddles were fashioned and Fred was finished and it looked good even though I say it myself, and Jumper and I had some good fun with it. Initially we had a bit of a stability problem in that it tended to capsize. After a bit of thinking and consultation with the Chippy we attached a couple of bilge keels and the problem was solved.

We proved how good it was by going on longer and longer voyages, with permission of course, and on one occasion ended up on some rich people's private beach but instead of booting us off we were invited to join the party they were having and were fascinated to learn about our home-made vessel. We eventually left having eaten our fill and a little the worse for wear booze-wise, but managed to somehow paddle our way back with no mishaps.

Interest increased amongst the crew and we eventually started to lend it out and it became a very popular recreational activity. Unfortunately on one occasion one of the pair that borrowed it on this occasion was very fair skinned and very susceptible to sunburn, and having suffered badly in the past, always covered up well. Unfortunately this particular time he wore one of the, then new, nylon shirts which, unbeknown to him, offered next to no protection. The pair had a good canoeing trip and on returning and re-stowing Fred, went below for a shower and the poor unfortunate on removing his shirt also removed most of the skin from his back that had stuck to it. He ended up in great pain and on a fizzer for negligence. The only real snag with Fred was that she weighed a ton mainly because of the heavy duty canvas used and the large quantity of paint we had used to make her watertight. We were ordered to stow it on the Boat Deck which was quite high up. It was not bad for launching but to haul it back up afterwards was more than hard work and a pain in the bum. I can't remember exactly what happened to it but I think we donated it to some children in Bahrain!!

Back Up

When the Stuff hit the fan with Iran and Iraq we were, as I've said before, at the mouth of the Shatt El-Arab ready to do the rescue bit.

I found out that a mate of mine, John Henry, H to his friends, had been drafted to the Gulf with a few others to take over an ex-RN LST (Landing Ship Tank) which had been loaned to a British petroleum company, and had been repossessed by the RN with a view to possible evacuation backup.

The reason him and his mates were chosen was because an LST had the same propulsion system as the Bay Class Frigate they had just brought back from South Africa. The layout being the same as a Loch and Bay Class Frigate, but the Bridge right back aft above the Boiler Rooms and the after bridge overlooking the Engine Room containing the same two, four cylinder triple expansion open reciprocating engines.

It appears that the civilian crew had not looked after her very well so the new crew had to do a lot of hard draft to get her shipshape again. One of the main jobs to be done was to get the bow door (cum-load-and-unload door) operational again because the civvies in their infinite wisdom had welded it up and the operating gear was disconnected and in very poor condition.

Eventually in record time the bow door was unglued and repaired, the operating gear was overhauled, reconnected, and after a few snags able to be operated again. Everyone including the skipper were pleased with the progress as it was an essential piece of equipment should the evacuation be necessary.

So it was then work-up time with the experienced few of H and his mates showing the new larkers how to do it. They ended up doing manoeuvres in and around the mooring jetties of Kuwait. The Skipper was a Lieutenant Commander who had not enjoyed mush sea time, and the rest of the engineering staff was not much better. So H ended up in charge of the Engine Room as he was the only one who had had anything to do with Recips and their auxiliaries. The so called Chief Tiff not having a clue, he said he was a turbine driven Destroyer man but in reality an ex-Barrack Station and about as useful as a chocolate fire guard.

As usual the Engine Room was boiling hot and H had the bearing cooling water rails used to good effect as a shower tunnel, and the skylights wide open.

The main manoeuvres that had to be performed were going alongside and driving up a beach and dropping the bow door. The beaching bit was easy as the Skipper had only to drive her in at ninety degrees to the beach, but coming alongside was a completely different kettle of fish, with lots and lots of engine telegraph orders and Main Engine revolutions, with the Engine Room crew going round like one-armed paperhangers, and the Boiler Room crew doing the same.

Eventually H got so worked up, frustrated and annoyed by all the Telegraph Orders just to come alongside, that he shouted out real loud in sheer anger: 'What the f—in' h— does the Old Man think he's got down here, a coach and four?'

Unfortunately, with the Engine Room skylights open, and the bridge wings very much above and in earshot the Skipper heard him and shouted down via a megaphone, very angrily in the negative. But it had the desired effect and the Gaffer eventually got the hang of it and finally signalled Their Lordships that his vessel was in all respects ready for action, and they were ordered to standby like us at the mouth of the Shatt el-Arab.

Armilla Patrol

Tiffs mostly stick together because other branches tend to be jealous of our relatively rapid promotion and resent us. Another reason is that on most RN vessels Tiffs have their own mess. Anyway I got drafted to another vessel and found out that one of my new messmates was someone I remembered from apprentice days. We ended up good mates and in course of time while nattering over a pint found out that in the distant past that we had both experiences the 'Armilla Patrol'. A pathetic attempt by the RN to police the Persian Gulf and general area. It appears that we, on our respective vessels relieved each other. He told me about the time of the first Iran-Iraq conflict where UK Nationals employed by the oil industry were at risk in the Basra area and his vessel was sent to the mouth of the Shatt el-Arab on immediate notice for steam ready to dash up river to carry out an evacuation if required.

The vessels we were manning were World War Two Frigates with pressurised Boiler Rooms, open front furnaces and twin triple expansion open reciprocating engines the whole thing being, at the best of time, very hot and heavy work.

The worst thing was that it was the time of the Shamals, the date ripening winds coming off the desert, the hottest part of the year. The Deck Apes sat around in the shade on the upper deck, fishing, dozing and generally lazing around, while the boiler and Engine Room gangs were on full watch keeping routine, with pressure up on both boilers and engines turned over every ten minutes, all sweating their little socks off. It caused more than a bit of animosity and a few near punch-ups. It was that hot down in the machinery spaces that two Stokers had eventually to be evacuated suffering from acute heat exhaustion, the rest of us just kept drinking gallons and eating loads of salt tablets.

The Engine Room gang had one trick up their sleeve to combat the heat. With open reciprocating engines if a bearing tends to overheat at sea water pressure main ran alongside each engine with outlet nozzles at closely spaced along the lengths. These nozzles were able be individually isolated, rotated and extend, and you've guessed it, all the nozzles on both sides were turned into the centre, fully extended, elevated and turned on to produce a salt water shower tunnel, cool man!!

That's Better

I was serving on an old Loch Class Frigate out in the Persian Gulf, my first seagoing commission. I was a very junior Tiff and was assistant some of the time to a senior Tiff who looked after the refrigeration and air conditioning systems. The systems had all sorts of problems not least of which was corrosion.

The cooling water for the refrigeration and air-conditioning units was sea water and its temperature could be as much as blood heat which wasn't much help. The cooling water pipework was standard mild steel which had seen better days and with the prevailing sea water conditions corrosion was rapid and as I've explained the system was nowhere near new.

Every now and again part of the cooling water main would spring a leak and the only way to deal with it was to isolate the leak, get the Chippy to build a wooden shuttering box round it and pour in a slurry mix of underwater cement. Fortunately it was very quick curing and the line could be opened back up and put on line and back in service within hours and the shuttering could be removed any time.

One continuous pain in the bum regarding this part of the job was the detachment of Royal Marines who were always complaining that their Mess was the hottest of all the Messes. We didn't dispute the fact but as we explained the reason was not so much the fault of the air conditioning unit but the fact that the Mess, as was traditional for the Marines, was in the Fo'c'sle with the Deckhead and both bulkheads fully exposed to the very hot weather. We tried all sorts of ideas to try and increase their air conditioning units efficiency to no avail. In desperation the Chief Tiff ordered us to shift the expansion valve of their unit up into the Mess so that the Bootnecks could watch the refrigerant expanding (which causes the cooling effect) through the sight glass of the valve, and also see the discharge pipe from it with a large lump of ice and frost on it. The refrigerant cooling effect was obviously no better and possibly slightly worse but the Bootnecks swore blind that it was much improved and no more complaints were received.

As a post script by the time we arrived back in Pompey almost the whole of the sea water services system pipework was concrete!

HMS INTRIM *LOCHINVAR*

Reclaim

HMS *Reclaim* was at the time the Navy's only Diving Vessel and getting to be a bit long in the tooth but still doing a good job. H and I were seconded to her as extra Engine Room Watch Keepers for a special Fleet Exercise as we had both served on and were both used to Triple Expansion Reciprocating Engine Propulsion.

The crew were a great bunch of blokes, especially those in the Chiefs and Petty Officers Mess and were very generous with the G at Tot Time, and also there was always loads of cans of ale, the only snag was that the food was atrocious. The reason was, that in the RN at the time, the various messes had a choice of catering method, either combined or canteen messing. Each crew member was allocated a fixed amount of money per day for catering which is multiplied by the number of mess members. With combined messing the menu and distribution of food is organised by the Catering Officer, but in canteen messing the Mess President is issued with the money and is supposed to buy in the ingredients from the Catering Officer, prepare the food in the mess and deliver it to the Galley for cooking. But in this case the Mess President, Chief Diver Pancho Powers, was always short of cash because it was mostly spent on beer!

While me and my oppo were serving temporarily on HMS *Reclaim* we met this Killick Diver, a bit of a character. He was a bit of a ladies' man, tall and good looking, they used to hang round him like flies round a lump of you know what. The type who always pulled when he went ashore which left everybody else feeling very inadequate, and hating him for it mainly because they didn't understand what he had that they didn't .

His party trick was to hang by his toes from a hammock bar. He only proviso for performing the said feat of athleticism was that someone had to help him down at the end of the show. He eventually left Reclaim waiting for another sea going draft, and the next time we saw him was when a towny of mine invited us up to his mess in Pompey Barracks, and who do you think we bumped into. You've guessed it, the aforementioned Killick, but with both arms in plaster, and I, like a fool, asked him what had happened and was of course told that he'd done his party trick, but that the rotten bastards who had encouraged him to show them his party trick had pissed off and left him dangling with the inevitable result. But as he explained it had its compensations because when he was first treated, and even afterwards in BMH he had all the nice female nurses to help him with his ablutions and bodily functions. Typical Jack, always remembering the good bits!!!

Players One

On return from our sojourns in the Persian Gulf and South Africa H and I ended up drafted to HMS *Adamant*, a Submarine Depot Ship. Soon after we joined it was off on a DF run to Portugal, a superb run ashore.

We were both laid on the beach at Estoril watching a couple of sailors who were Shallow Water Divers who, having smuggled fins out of the diving store, were showing off in front of the local talent, and very nice they were too. There and then we decided to slap in for a shallow water diving course. That was a bad mistake!

When we got back to UK our Draft Chits to the Diving School at Rosyth Dockyard had arrived, in a very cold very snowy February. By now we were Third Class Tiffs entitled to wear three gold buttons on each sleeve cuff and be addressed as Chief, all the rest of the course members were mainly ODs, and Killicks so we thought we would be treated well, but not a bit of it. It was every man for himself, we did our dips off the deck of an MFV in the Tug basin of Rosyth Dockyard, having to sweep the snow off the deck beforehand. Then it was a mad scramble to get a decent suit. Some of the suits were a disgrace, patched up with bicycle puncture repair patches. The next problem you had was to get into the dammed thing. You first put on thermal underwear, which in those days was abwool which was very itchy and also very absorbent. Once you had the jersey, socks and long johns on you climbed into the dry suit through the neck hole, until you're in it up to your waist, then you need help. Someone has to grab a sleeve and the neck hole on the same side and in a co-ordinated opposite lunge the prospective wearer has to plunge his arm into the gaping neck hole and hit the sleeve hole or else. Then you have to perform the same manoeuvre for the other arm, then it's the dreaded neck band that's fitted inside the suit neck hole, then the hood is pulled over the head the skirt of which is arranged over the now expanded neck hole. A large jubilee type clip is then slotted over the head and tightened over the sandwich of neck band, suit neck and hood skirt. If you think that's complicated you ain't seen nuthin' yit. It's the breathing set next, an oxygen rebreathing set with high pressure oxygen storage cylinders on the back with an envelope arrangement underneath which contains a quantity of four ounce lead balls to assist you in sinking and a quick release pin on the envelope to jettison the contents to assist in surfacing in an emergency.

Strapped to the front is a breathing bag with a large canister connected to it which contains soda lime which absorbs carbon dioxide, the poisonous gas that we exhale. This is why it's called a rebreathing set because you breathe in oxygen from the cylinders on your back via a reducing or demand valve, breathing bag and soda lime canister. Exhaled unused oxygen is stored back into the breathing bag and the generated carbon dioxide is absorbed by the soda lime.

The reason for this seemingly complex system is to prevent discharging any give away bubbles when you're on covert operations, so when you watch a film with frogmen in it inspecting enemy beaches or shipping and streams of bubbles keep appearing, you know they're breathing air and the film's rubbish!

The diving course itself was truly terrible but it did have its lighter moments. The accommodation was disgraceful, it was an unlined wartime nissen hut, the only heating being an old cast iron pot bellied stove, the fuel, coke. Have you ever tried lighting that sort of gadget in the freezing cold with soaking wet coke. It's easy, you nick a full breathing set cylinder, a bit of newspaper on the grate, a few sticks on top, then fill the whole thing with soggy coke, light the paper and give the whole thing a steady blast of the old oxygen it burns up like rocket fuel.

But I digress, to recap, we have the suit and breathing set on, next comes the dreaded wrist bands which are just very wide lazzy bands, the idea being to prevent ingress of oggin up the cuff of the suit. In actual fact all they do is cut off circulation to the hands which in extreme cold ensures that all feeling to the digits is completely lost. Next comes the even greater dreaded nose clip which is constructed of two half-inch diameter rubber pads connected together by a looped coil spring. The two pads are prised apart and placed over the flare of the nostrils and released thus closing the nostrils rather firmly so preventing water ingress to the nasal passage but which, after a time tends to bring tears to the eyes. The snag with me is that I'm rather well endowed in the bugle department and due to that had to have two springs which, after a while, felt like I had a G clamp attached to my face. Now the mask, you have to gob on it, on the inside, and rub it in, it's supposed to stop it steaming up, not that it would make a difference in the Tug Basin, you couldn't see your hand in front of your face.

Next you evacuate the breathing system of air by use of the way cock which is attached to the mouthpiece front. The object of this gadget is to evacuate the lungs and breathing set of air so that the soda lime is not contaminated, and unwanted nitrogen is eliminated. The method is – way cock to atmosphere – exhale – way cock to oxygen system – inhale deeply – repeat at least three times. I forgot to mention that you are now wearing either fins for swim work or heavy boots and lead-loaded belt for bottom work. A lifeline is attached to your harness, the quick release buckle on your harness is checked and, with a tap on the head you put your hand over your mask and jump, that's all there is to it! Your next problem is sinking, you're floating like a beached whale, but all is not lost, on the back of your hood is a bit sticking out that looks like a flattened inflation nozzle of a balloon. It is in fact a non return valve, and the reason you're floating is that all that abwool has air trapped in it, so now you have to tip your head forward and the hood valve allows air to escape then you sink. You're now in another alien world of no sight and very little sound, apart from the thrashing of Tug propellers, which totally ignore the 'Diver in the water' signal flag.

Fender

The shallow water diving course was not a lot of fun because of the weather and a lot of the course participants had chickened out through 'heavy colds' or so they said, but in reality it was getting a bit tough and our instructor didn't take prisoners.

The classroom work turned out to be a bit of a struggle for a few others, even though H and I tried to help them out they fell by the wayside.

We were by this time about half way through the course, and during a classroom session one of class asked our instructor 'Pancho' Powers why we weren't issued with divers' knives, very large, very sharp, double edged, non magnetic instruments anymore. The surprising reply was that one of his earlier students had been attacked by a particularly vicious 'Man Eating Pudding Fender' and had almost cut himself out of his suit and had also received some rather nasty self inflicted wounds.

The class as a whole were amused but a bit confused so Pancho explained. A pudding fender is made of what looks like knitted rope in the shape of a sphere, hence the name pudding, the filling used being scrap and offcuts of grass, a type of rope that floats. It has a securing rope attached to it which sinks.

When a fender is damaged it is usually just dumped over the side, it becomes waterlogged and sinks because the securing rope drags it down, but only till enough of the said rope is on the bottom when the buoyancy of the said fender takes over and it ends up lurking in suspension. The poor unfortunate diver, in zero visibility, bumped into it, which unnerved him, what he didn't realise was that it was on its way back to him but with a bit of velocity. When it hit him back it was out knife and slash and stab, with disastrous results, and the banning of divers' knives for trainees.

Diving Course (Rosyth)

The bloke in charge of the whole training programme was Chief Diver Sidney 'Pancho' Powers. Pancho was a great bloke, but a stickler for accuracy and detail. His instructions to you on the start of a Dip might be to head out to a floating fag packet and retrieve it without being seen. It sounds easy, but in reality, with no point of reference, you lose all sense of direction and eventually you have to come up for a shufty to get your bearings. Beware showing too much above the surface, 'cos the next thing you know is Pancho pelting you with 4 oz lead balls very accurately and they don't half hurt.

Working on the bottom in lead boots and belt is an entirely different kettle of fish. A 'Shot' line consisting usually of a concrete block weighing a stone or so with a metal ring cast into it with a two inch or so rope attached to it. This is dropped over the side and secured to a bollard on the deck of the MFV. You are then dropped over after it equipped with a light half inch lead line. The idea is that when you reach the bottom you attach your lead line to the shot line and walk away from it whilst paying out the lead line (so that you can find your way back). That's the theory – in reality, once you hit the bottom you are thigh deep in thick glutinous mud. Having attached your lead line, to try to move you have to lean forward almost horizontally and start paddling in the mud, paying out the lead line as you go. You can't guess your actual progress, but you feel you're doing real well and when all the lead line is paid out you're pretty exhausted. You stand up and turn round to go back and hit the shot line with your lead line coiled at your feet.

To add to your misery, whether swimming or on the bottom the cold and leaks get to you. The cold takes its toll particularly on your hands as they are totally exposed and as previously explained, the pressure of the suit cuffs and wrist bands reduces circulation badly. But by far the worst were the leaky suits. You start the dip and slowly this damp feeling starts and gets slowly worse and colder. But by far the worst snag with leaks though are that they make you get heavier and heavier and you have to keep opening the regulator bypass to allow a bit more buoyancy into the breathing set, but there are limits. The heavier you get the harder it is to work with progressively increased exhaustion.

The other thing you have to be careful of is a 'cocktail' which is caused by salt water leaking into the soda lime canister as this causes a violent chemical reaction. This nasty mixture then erupts through the mouthpiece with no warning with extremely horrible results.

Regarding leaks, one amusing incident involved H. As usual we were waiting for the signal to go below and get suited up. H had pulled flanker and manoeuvred his way to be nearest the companionway. The order was eventually given and H was down below like a rabbit down its burrow and lo and behold he had chosen a beautiful chalky white suit, it must have been brand spanking new. We all got suited up and rigged in our sets etcetera. H went in first in his brand new suit, sank like a stone and immediately shot back up to the access ladder protesting violently with gestures. But Pancho would have none of it and kicked him back down and told him to get on with it. By the end of the dip a very sad and very sorry and very very cold H emerged. It transpired that the suit he had chosen was a very old but unused type which was rigged with a Pee hole and a blanking plug on a short lanyard. Unfortunately on this unit the lanyard had broken and the plug was lost and H hadn't noticed. As he entered the water an awful lot of very very cold oggin hit his nether regions, which is why he tried to exit like a rocket. So at the end of his dip poor old H's legs were full to the brim and the water had capillaried up to his armpits as abwool acts like a very good wick. He just about managed to drag himself out weighing a ton, convinced that his sex life was ruined. After that he never failed to check any suit he chose thoroughly before getting dressed.

With visibility in the Tug Basin zero, to check our technique we did dips in the Olympic swimming pool at HMS *Caledonia*, the Shore Establishment H and I had completed our Apprenticeship in. The good thing about this was you didn't have to prat about with a suit, the only snag was the pool itself, being of Olympic standard it had the normal shallow end and a very deep deep end. The problem was that Pancho insisted that in swimming lengths of the bath you hugged the bottom with the inevitable change in buoyancy. When going from shallow to deep you lose buoyancy because of the compression of oxygen in the breathing bag of the set. This compression can be compensate for by cracking the bypass valve of the regulator for just a little burst to increase buoyancy, but beware the return trip because if you've overdone it, when you come back up the slope towards the shallow end oxygen starts to leak out of the sides of the mask, or even worse you end up surfaced. The next thing you know, Pancho is either heaving 4oz lead balls at you, or jacking you in the ribs with a very long bamboo pole whilst shouting at the top of his voice 'OXYGEN WASTER'.

Another amusing incident involving H was during one of these training sessions. To amuse ourselves, and with Pancho's blessing, we used to hold races to see who could get into the water first. Our sets were lined up along the deep end wall. We would line up at the waters edge facing our own sets. On the command you dashed to get your set on, clear the air from the set with three out and in cycles. On this occasion H was first in, but shot straight out again coughing and spluttering because, in his rush to win he had miscounted and left the waycock in the atmosphere position and guffed up half the pool.

Other techniques practised were changing sets under water and rescuing somebody from the bottom by sharing one set, with the ever vigilant Pancho at the ready with bamboo pole and the lead balls.

Throughout the course classroom work was very important, as was the workshop practice where we serviced our sets and pumped up our set cylinders from larger storage cylinders. H and I had no real trouble with either element, but some of the lads struggled and even with our help eventually dropped out. We also kept losing men through illness, mainly heavy colds, until eventually we were the only two left. He wouldn't give in 'cos I wouldn't and vice versa, the main reason being that we were nearly the end of the course. The final irony was that one of the last elements of the course was the decompression chamber dip on the next Monday.

On the Friday before our scheduled dip Pancho took us into a deserted area of the dockyard and there in all its glory on a shingle covered weed infested patch was a dirty rusty skid mounted cylinder called a Decompression Chamber connect to an engine driven compressor. Pancho suggested that, as we were Tiffs we should check things over. On inspection we discovered that it was a bit unusual in that it was dual fuel, starting on petrol and changing over to diesel when warmed up. The starter batter was as flat as a fart, so we drew a fully charged unit out of the Dockyard Stores on Pancho's slop chit and tried again with no joy. By this time we were getting very suspicious of Pancho's motives! Doing a compression check (to check piston and valve condition) we found that three of the four cylinders were well below acceptable values, and no wonder, H had found traces of charred rag in the air intake. Some silly sod had obviously tried starting the thing on diesel using an old and very frowned on trick of removing the air intake filter and holding a burning oil soaked rag near the air intake whilst turning the engine over so preheating the combustion air. Unfortunately whoever did it got the burning rag too close and it got sucked in. We buttonholed Pancho but he denied any prior knowledge but we didn't believe him, but when we explained our fears that it might have bent valves, his uncompromising answer was that if it couldn't be fixed the course couldn't be completed. H and I went into a huddle and decided that as we'd come this far it would be daft to give in at the stage we were as so we decided to give up our weekend to see if we could fix it. By the time we'd made this decision it was getting dark, there was no lighting around, no power points and we tried but failed to borrow a generator so it was going to have to be daylight.

So, early on Saturday morning we bummed some tools and started to do an engine top overhaul. On removing the cylinder head we found charred rag under three inlet valves, and out of four you don't stand any chance of running it up. On stripping the head down we found no bent inlet valves but two burnt exhaust valves that were nothing to do with the rag job but still needed attending to if we were to stand any chance of success. These valves needed precision grinding so we decided to bum off 'Cally' (HMS *Caledonia*) our old stomping ground. We found the workshop Regulating Chief Stoker, and H sweet talked him into opening up the tool room for us and gave us permission to use the valve grinder, but it was designed to grind larger valves so I turned to and made a centring sleeve. Eventually we had everything ready to reassemble before we lost the light. Sunday morning dawned bright and cold, and we got down to it. We begged borrowed and stole grinding paste for lapping valves in, jointing compound for the rocker cover etc., packing for the cylinder head joint and feeler gauges for setting tappets. Having finally got it all reassemble we checked petrol, diesel fuel (having first drained off nearly half a pint of water from the tank), cooling water and oil for the engine and oil for the compressor, and both air intake filters. Came the moment of truth just before dusk, we tossed for it and I won so H pressed the tit which caused just a whirring noise and nothing else. He tried again with the battery slowly giving out when suddenly a couple of smoke rings puffed skywards, and then 'Verroom' and it was away. The governor took over and it warmed up nicely and was then switched over to diesel. It gave a couple of coughs and then came back on the governor. The compressor clutch was engaged and a satisfying hiss emanated from the open hatch of the pot. We shut it down and shook hands, all we needed then was for the cylinder head joint to last out till after the next days pot dip. We cleared away the debris, got a quick wash and brush up and thumbed a lift into 'Dumps' (Dunfermline) our old apprentice days stomping ground, and got pissed.

Came the dawn, a bit the worse for wear, Pancho came into the classroom looking quite glum and asked how we'd got on and was gobsmacked when we told him it was fixed.

We did the classroom bit about procedure and safety and then went to the tank to do the 'Dip'. We started the engine no trouble, then climbed in to the pot, Pancho shut the hatch and engaged the compressor clutch and slowly raise the pressure to represent the equivalent of 100ft depth, equivalent to three atmospheres, stayed there for the regulation 10 minutes and then slowly decompressed. Due to the previous nights run ashore we were not in very good condition but survived and on return to the diving school had a couple of so guffs of oxygen and felt a lot better, and we eventually finished the course. We were awarded our shallow water diving badges of a gold diver's Hard Hat and diver's log book, sewed the badge on the right sleeve above the three buttons and felt very pleased with ourselves.

We were both then drafted back to *Adamant* to work in the workshops, I in the light machine shop, H in the heavy machine shop. Depot Ships are amazing vessels fully equipped to cope with any job and any emergency.

The snag with all this expertise and up to the minute equipment, when things were slack, which was often 'Guvvy' reigned supreme. You could get anything (for a price) cares being one of the best money or tot spinners. New exhaust systems, accident repairs, chassis welding, you name it, it could be done.

Testing Testing

Chief Diver Powers, Pancho to his friends, was serving on HMS *Reclaim* the RN's main Diving vessel at the time of the incident. During stand easy one day while Reclaim was in Dockyard hands, undergoing a minor refit, he asked his PO Diver if he'd done the regulation pressure test on the main Decompression Chamber. He got a negative answer, an apology, and the offender shot off to do as he had earlier been instructed, and returned to the mess in record time to finish his cuppa.

This type of pressure chamber is a most important piece of equipment, having saved many lives and prevented crippling of Divers from the bends due to too rapid ascent from long deep dives.

Anyway stand easy was almost over when there was a loud knocking on the mess door, it burst open and a breathless Dockyard Matey blurted out that his oppo was trapped in the 'tank'.

'What Tank?'

'That round one with the portholes!'

'THE POT!!!!!'

Pancho grabbed the PO and demanded to know if he'd done the proper regulation Checks inside the Pot before he'd closed the hatch and applied the test pressure and was told in a very apologetic manner that he'd just slammed the door shut and wanged the pressure on 'cos his tea was getting cold.

What had happened was that our Dockyard employee, just before stand easy, had espied a nice long upholstered seat in a vessel through an open door. So when Stand easy was piped, quick as a flash he was in there for a few crafty zeds. The next thing he knew was the hatch slamming and a tremendous increase in pressure which gave him a very nasty headache and burst eardrums.

When they got to the Diving Flat and the Pot and peered in through the observation port Pancho reckoned that if they'd left him in there a few minutes longer he'd have had all the paint off the inside of the vessel because he'd obviously tried to use the telephone and, when he got no reply, had ripped it off the bulkhead and used it as a hammer to try and call attention to his plight to no avail until his mate had peered in seen his problem and raised the alarm.

The PO looking rather guilty, went to decrease the pressure by the book, gradually, but Pancho stopped him and dragged him to one side and explained:-

'The pressure in there is one and a half times its maximum working pressure. Him in there is in great pain and he's also very angry and confused and if you ease the pressure off slowly and open the door nicely for him he's going to, at the very least kill you.'

'So, do it this way, you stand by the hatch with his mate, and when I say now open the hatch and get his mate to drag him out OK?'

With that Pancho just spun the vent valve wide open, the pressure fell like a stone producing a very heavy mist, and –

'NOW.'

The hatch was flung open, Pancho and his PO beat a hasty retreat leaving the poor dazed sod to be dragged out of the fog by his mate to seek medical assistance.

As Pancho said afterwards, philosophically about both parties.

'Look before you leap!!!!'

The Special

The three of us were lumbered in HMS *Lochinvar* South Queensferry. That was Danny, Mike and Me. Our daily routine was so boring, reveille, quicky breakfast then down to the jetty to hop on to an MFV and across the Forth to Rosyth Dockyard and get lumbered for any rubbish job we were given. Then at day's end back the same way to Lockinvar, into the canteen to drown our sorrows.

That is until one day Danny came across what looked like an old abandoned Hillman Minx at the back of the disused boat sheds. So Danny conned Mike and I into building a 'Special', but first he said we'd have to make sure the engine was a runner and the chassis was in good nick, this model was built in the days before monocoque construction. So we ripped the shot body off it and the chassis looked in good nick apart from a few minor dodge bits which would be easy to repair. But the engine was a different kettle of fish. We rigged up a temporary petrol tank and batter, but it took a lot of hard work and ingenuity to at last get it to fire up, minus exhaust. What a row, so we shut it down fast so as not to arouse interest in any nosy sods.

Danny then sent away to some southern firms who specialised in fibreglass body shells, and we chose a rather fetching bright red sports job. The snag was that they wouldn't deliver so we had to club together to hire a car and collect it from the south side of London and we only had from Friday evening to early Monday morning, and all this before modern motorways.

The hire car was a beat-up old Ford Consul, the cheapest we could find in Burgh, and we packed it with loads of cordage, sacking and sponge rubber. The week we chose to go the weather was atrocious with rain and fog all the way to Smoke, and only the nearside headlight worked, the wipers had seen better days, and the brakes were as handy as toothache. But apart from that it was a 'lovely' trip down!

We had a few near misses on the way down, but arrived in Smoke in thick fog and dark early morning so decided to park up safely and go and find a cafe and have a well earned breakfast. We had just about had our fill and finishing off a nice sweet mug of tea, when in strode this Bluebottle, and enquired if anybody owned a beat-up Cortina. We admitted to same so the Law said in a loud voice, 'Remove it because it's causing an obstruction!' So we trooped out to a fine dry bright morning to be shown our lump of transport parked in the middle of a very busy roundabout with lots of rubberneckers enjoying the spectacle, and we, red-faced, climbed aboard, with Mr Plod holding up all the traffic allowing us to go on our merry way.

We eventually found the Body Shell joint and were shown our choice of colour and style and it looked superb! We paid the man and lugged our purchase onto the roof of the Cortina having padded it with the sacking a sponge rubber then lashed the whole lot down with the ropes going through the Cortina's windows with cloth and bummed sheet metal protecting the fragile edges of the shell.

We were ready to roll around noon but hit bad traffic going back through Smoke, and it took forever to get rolling properly and it was touch and go whether we would make it back in time. It was Danny's turn to drive, with no licence and he had one really bad habit: he didn't like using the brake pedal! This produced some very hairy moments as he ended up nearly going up the arse of quite a few very angry motorists. The final straw of this technique was a very irate articulated wagon driver who suddenly anchored up to a stop and leapt out of his cab and marched purposefully towards us, so Danny, ever ready to have a punch-up, bailed out of the car, bent down to clear the rear of the trailer and waited for the confrontation which didn't materialise because when the wagon driver saw Danny uncoil himself and saw how large he was, issued a few expletives and dashed back to his cab and shot off at a rate of knots, and Mike and I heaved sighs of relief, Danny remounted and off we went again.

Eventually dusk turned to night and Mike took over the driving much to my relief and we were making good progress and I was dozing off in the back when suddenly with a screech of brakes, I was thrown forward (no seat belts) and smacked my head on something hard which knocked me out for a while. What had happened was that the sacking, due, we think, to Danny's erratic driving, had worked its way forward and with a slight change of wind direction, had flopped over the windscreen, leaving Mike completely blind! It was a good job nothing was close behind.

With the problem sorted we went on our merry way with me doing my stint of driving. We eventually reached the outskirts of Edinburgh and Mike took over again as he knew his way round the area. Eventually as dawn was breaking we arrived at *Lochinvar*'s main gate, and the duty Crusher flung the gates wide open and waved us through. Mike wound his window down, expecting some abuse, but instead he was told to get inside 'and hide that heap of junk'. Mike took no further bidding and shot off and parked behind the sheds in a corner. We'd missed breakfast and had to shoot off for the MFV.

We were all completely knackered but well pleased that we'd made it, but the final irony was that we'd been ushered through in that fashion was because some big wigs were due any minute and the Chief Crusher (MAA) didn't want us anywhere near!

By the time we got back to *Lochinvar* we were out on our feet but we still had to unshackle the shell and get the Cortina back so I drove and Danny followed on his Triumph Thunderbird. That was a mistake, I should have let Mike take my place, but we got rid of the car and paid the man and told him it was a bag of tish and not to expect us back, then I put the spare helmet on and mounted the beast behind Danny. The next thing I remember is I'm sat on my arse in the middle of the road 'cos he'd taken off like a bat out of hell before I'd had time to hang on. The problem was that he didn't realise for ages that he'd lost me, and there's me standing like a wally on the pavement edge, complete with helmet.

Eventually he came back for me and as I remounted I took a firm hold of him so that if I went this time he'd come with me. It was one of the most hair raising experiences I've ever had and it frightened the life out of me. We wove in and out of traffic, completely ignoring speed limits, turning corners at alarming angles. I just hung on, closed my eyes and prayed.

When we eventually arrived back at *Lochinvar* in one piece and parked up I asked him why the big rush; the reply was he just wanted to get back before the bar shut. So we joined Mike in said bar, and sloshed and died until reveille.

Our evenings were now spent tinkering with the Hillman chassis and only getting to the bar for the last couple before last orders. Our trips to the Dockyard also proved fruitful because we had befriended a few tin bashers and squirrelled away aluminium checker plate, plain aluminium sheet and pop rivets by the bucketful. But the weekends were where the major progress was made. We had the shell mounted on the chassis and a lot of the shell interior was taking shape. The rolling gear was up and running, the steering was connected up and adjusted, and the brakes shackled up and tested, and we were about to start the lights, dashboard and prop shaft when suddenly out of the blue all three of us got drafted to Faslane, the Submarine base above the Clyde!

What to do? By now Mike had bought himself a little Ford Prefect so it was decided that we would load all our remaining ill gotten gains, and also our own personal gear onto both vehicles and tow the Special across with the Prefect, but the burning question was, who was going to 'aim' the Special? We drew straws for the privilege and guess who won, or maybe it should be lost? ME. I'm sure it was fixed but couldn't prove a thing.

So on the appointed day we had everything ready, and I was togged out in a foul weather top, Danny's spare helmet and a pair of Mk 21 goggles and looked and felt like a prat!

The tow all started off a bit hit and miss like a seesaw until we got the hang of it and became more confident. It being a pleasant day I started to enjoy myself a bit and we eventually made Stirling and stopped for a quick breather. The next bit should have been a doddle as it was 'the' Stirling straight all the way to the Loch Lomond Arms Hotel. That's where things went all pear shaped because unbeknown to me Danny had talked Mike into letting him have a spell at driving. Off we set again but suddenly instead of the long straight road I'd looked forward to Danny took to the hills. It wasn't too bad for me going uphill as the little Prefect puffed and panted, but going downhill was a nightmare 'cos Danny just put his foot down and we flew. I tried to slow him down by braking but I could smell the drums getting too hot. It was just another hair raising experience kind permission of Danny, and we eventually came back onto the planned road at the Loch Lomond hotel so it wasn't quite so bad until we were approaching Helensburgh where it started to rain. My goggles were as useful as an ashtray on a motorbike and coming down the steep hill into the centre of Helensburgh in the rain was another nightmare. I think I had all four wheels locked up most of the descent and more or less slithered to the bottom. We then turned right towards Faslane and stopped at the nearest Hostelry, which I can't remember the name of and had a few wets when Danny got to his hind legs gave us both a wink and went up to the bar and spoke to the Landlord who got all irate and told us all that the bar was shut and to clear off out of his boozer. Danny was laughing his hat off so when he'd calmed down he explained that he had noticed that there was a sign over the bar boasting that they stocked every known brand of Scotch whisky, so he'd asked him for a Jameson's. So after a good laugh we wended our weary way past Rhu to Faslane compound, where we were all three relieved to still be in one piece!!

I soon got a draft with Mike down to the HMS *Dolphin* for the Submarine course while Danny joined the repair staff of HMS *Adamant* the Submarine Depot Ship. I never did find out what happened to our 'Special', but I hope it went to a good home of an enthusiastic mechanic or that big car graveyard in the sky, Amen!!

Players Two

Most branches in the Royal Navy, if they were honest, deep down hate 'Tiffs' because of their rapid advancement, but they all knew where to go for help when they were in the shit!

Even the Master at Arms or Jaunty (The Senior Regulating CPO), God to most, asked me if I would have a look at his car. It was an old sit-up-and-beg Rover 90. Just the tappets and a bit of an exhaust blow, he said! It turned out that H and I spent all weekend again. The tappets were overhead inlet and side exhaust valves, a bugger of a job with no room to work. The exhaust was even worse, a complete new system was required, which passed through the chassis members and used triangular connecting flanges, a right sod of a job. We had to use a lot of 'on' to get it ready to fit.

But in the long run these sort of jobs pay dividends, as in this case. One of our number in the Tiffys mess had transgressed and was in the rattle. Some time earlier he had done a guvvy for the Admiral's Chief Steward. Unfortunately he was busy when the Tiff went to collect his dues but rather than turn the poor lad away he sat him down and in the stewards' pantry, plonked a large bottle of gin and a bottle of orange cordial and a glass in front of him and left him to it. David 'Tugg' Wilson (all Wilsons are Tugg in the mob) didn't have the heart to tell him that he didn't like gin. 'Still,' he thought, 'I'll have one to show there's no hard feelings.' Unfortunately he began to quite like it, eventually when the bottles were empty he decided to sod off and was last seen by a particularly snide junior member of the regulating branch staggering into the heads, where from disgusting noises emanated. Fortunately one of our mess stewards saw said RPO legging it to the Jaunty's office, and told us what he'd seen.

H decided to see if having a word with the Jaunty himself would help as being drunk onboard was a serious offence. We don't know what H said, and he wouldn't tell us, but whatever he said it worked. Tugg got a right bollocking but nothing more. The snag was that if he'd been trooped the Steward would have been implicated and also trooped, so H got a good result.

We, that is H and I, eventually got very bored with the general day to day routine and both volunteered for Submarines! It seemed to be a good idea at the time. So we both slapped in (requested in writing) to join Boats. To my dismay H's request was granted, and he was told he would be drafted in time for the next Submarine course, whilst my request was refused! I was gobsmacked, and slapped in through my Divisional Officer to question why I had been turned down. My DO eventually came back to me, and in an embarrassed manner, told me he had made me an appointment with the base MO. He would not be drawn on the reason, so I had no option but to wait until my appointment came up. The consultation was short and sweet, the MO, a Surgeon Commander, came straight to the point and told me that I had been refused my request on the grounds of my habitually contracting sexually transmitted diseases. I was absolutely floored, as although I'd had my moments nothing like that had ever befallen me. I protested and asked to see the file he had been consulting. He was good enough, as he was not obliged to do so, to show it to me. Waves of relief passed over me as the file was not mine, but someone with a similar name. To cut a long story short the correct file was found, my request was granted and we both went down to HMS *Dolphin* together. The Sick Bay Tiff (Sick Berth Attendant) was given a mild bollocking for getting the wrong file, so no real harm was done, and honour restored!

The Boats course was great, the technicalities were no problem for either of us. We had a great set of course mates who all helped each other, and of course H and I also combined our efforts which made things even better and a lot easier.

The Tank

The Tank was something else, the object of the course being to train a boat's crew to escape from a sunken vessel, in this case from 100 feet depth.

The first thing that has to be done is to check your buoyancy! Most people are naturally positively buoyant, but a small minority are big boned and just plain sink. Guess who sank, you've got it, H, so for the rest of the practical part of the course he had to wear a red hat. The course was good and the training excellent, and in those days the escape method was fee ascent, which means you used your natural buoyancy to rise 100 feet.

When you are at a depth of 33 feet one extra atmosphere (approximately 15 psi) is pressing on you but you don't notice it because your lungs are at the same pressure. At 99 feet you are at 3 atmospheres but again it is equalised internally. The snag comes when you depressurise when ascending because the air in your lungs expands and the body's natural reaction when underwater is to hold on to it, which will over pressure and burst the lungs. So one of the first things you are taught is that you must force yourself to breathe out for the whole of the 100 foot ascent. It takes an average 22 seconds to rise 100 feet, you try blowing out hard for that length of time!

It's quite clever really, the Tank is about 20 feet in diameter and 115 feet tall, the extra bit being a mock up of the interior of a typical Boat with a hatch in the middle of the deckhead. A small group at a time are herded into the mock-up dressed in cozzies, and in H's case, a daft red hat. Each one carries a demand valve on a short length of hose with a male instantaneous coupling, known as a PLC. The entrance watertight door is shut and sealed behind you, the instructor then moves you up to the overhead hatch and removes two semicircular wooden covers from a recess around the hatch and a heavy canvas tube known as the twill trunk drops down. It is secured at the top round the hatch with a water tight joint, and the lower end is kept in shape by a steel ring sewn into the skirt and ends up 30 inches above deck level and is secured rigidly by four light lines to ring bolts set into the deck.

Now comes the tricky bit, the compartment is flooded, which is when we plugged our demand valves into the PLC sockets which ran in parallel rows in the deckhead. As the compartment flooded and the water rose so did the pressure, and with the dull emergency lighting, a heavy mist formed to give a very eerie atmospheric feeling. With the water up to our chests the pressure equalised and our instructor who has an extra long air hose ducked under the rim of the twill trunk, climbed into the trapped air pocked and released the clips of the hatch which flew open and released the trapped air. The instructor then returned back into the compartment to supervise our exit. It sounds easy but it isn't that simple. First you have to shuffle along changing PLC sockets as you go, then it's your turn and you take a last breath, discard your demand valve, duck under the twill trunk rim and up you go forgetting all your training about breathing out, but suddenly you are grabbed and thumped in the gut, which makes you breath out rather explosively. What happens is that a Swim Boy has spotted your omission and has swum out from his alcove in which he has a refreshable air supply and given you a sharp reminder, and then popped back to his billet to peer out of his sight glass to look for his next victim. You're still not out of the woods though because the human body is less than aerodynamic and tends to shoot off at angles rather than straight up so the Swim Boys pop out spin you round and send you on your zig zag way. You eventually end up bobbing about on the surface, very disorientated and very relieved, but also quite elated at having done The Tank!!

Sea Trip 1

The Submarine course was great, the technicalities were no problem for H and I. We had a great set of course mates and we all helped each other, and of course H and I also combined our efforts, which made things even better and a lot easier.

Three things of this period stick out in my memory, the Canadians who trained with us, the first sea trip in a Boat, and The Tank!

The Canadians were great blokes who integrated so well, but couldn't get their heads round the English weather. Jim Matiatchuk from Saskatoon, Saskatchewan, was heard to utter in a loud despairing voice, having experienced, in a building with a corrugated iron roof, spring sunshine, heavy rain, and snow.

'God damn it, all we need now is hailstones and we've had the whole damned caboodle,' and the heavens obliged with deafening results.

The sea trip! To me it was everything I'd volunteered for. We were split into small groups and at various times joined a Boat and went on a day trip. Our ride was an old weary, but very happy HMSm *Trite*, an old wartime 'T' Boat with a good record in the Mediterranean theatre. She had a riveted hull as opposed to modern welding, and was still fitted with a four-inch gun sited forrard of the conning tower.

Visitors to Boats for sea trips are limited in numbers per boat because of both the amount of room available, but mainly because of the number of escape equipment sets that are carried aboard. We guests, huddled in the control room, could hear all the orders coming down the Bridge and also the responses. I was amazed at how calm the crew were at going about their business. A message came over the tannoy to 'open up for diving' and the crew still went calmly on. Eventually the klaxon sounded, down came the bridge crew, the conning tower lid was shut and then what appeared to be nothing else, no dramatics, nothing. The Officer of the Watch ordered open Main Vents, the Tiff on the panel operated some levers, and that seemed to be it. The Officer of the Watch seemed preoccupied with the instrument panel between the planesmen, so I looked in the same place and saw the depth gauge and realised it was moving, we were going down but there was no sensation of it. The Captain ordered the rear (search) periscope raised, then with a quiet swish the mast slid up and the ocular box appeared. The Captain peered into it and swung it round once and then invited us visitors to have a look. To my amazement and delight the view was incredible, so clear, and being able to switch to high power was better than high magnification binoculars, it was also able to elevate and depress and being binocular, object ranges could be measured.

We were then given a Cook's tour from stem (torpedo compartment, fore ends) to stern (motor room, after ends), the complexity of which overwhelmed us all. After the official tour we were given leave to do our own thing for half an hour. A few of us ended up talking to the Chef, a rather laconic Glaswegian whose work place, the Galley, was an alcove off the main gangway. How he managed to cater for the whole crew in such a confined space was a matter of wonder, which led to one of us asking him if he had ever had any hairy moments? He started chuckling, and eventually explained in a broad Glaswegian accent:

'Ah rememba uonce ah was dooin dinnah for the wardroom an some verra verra important vizators. It were rroast poork an all the trimmins, and ah were jist aboot ta dish up an a'd forgot the green beenz, so ah grabbed a tin o em an banged e int oven. One thing a'd forgot te doo was te peirs't tin an carrid on dishin up. Suddenly there were a hell o a bang, and ah ended up wearin the oven door an there were green beenz evera were, what a bloody mess. Everathing wer covered in f—ing green beenz, an a wer charged wi abuse o goverment property, but the Skipper saw the funny side o it an let me off wi a bollakin.'

Sea Trip 2

The tannoy on HMSm *Trite* called us visitors back to the control room, and the Skipper told us that his crew, for our benefit, was going to demonstrate a 'gun action' exercise, and warned us not to get in the way.

Action stations was called, and a few personnel changed places. The Skipper called for periscope depth and did a quick all round sweep to check that nothing was in the way, and ordered 'Stand by for Gun Action' and suddenly the control room was full of a very stern looking gun crew, all in wet weather gear? (the weather up top was perfect!). The gun layer, breech block slung on a lanyard round his neck, stationed directly below the gun hatch. The next order gave serious concern to us uninitiated.

'Pressurised the Boat,' and the diving panel operator, who was now the Outside ERA (Wrecker) opened a compressed air valve to atmosphere and our ears registered two senses, the terrific noise and the increased pressure, like going downhill very fast. The Wrecker was checking the atmospheric pressure gauge, zero being normal, and either side being plus for pressure and minus being vacuum. When the pressure reached the gauge maximum the Wrecker stopped blowing and reported the Boat pressurised. The next set of orders came thick and fast, 'Down periscope, increase to 140 revolutions, blow main ballast, planes to dive!' I found out later that the idea was to surface as quickly as possible and get off the first round as soon as possible. So blowing main ballast to give positive buoyancy, planes to dive counteracts the first action and increased speed helps. The next thing to look for is the tendency for the depth gauge to reduce, at which time there is another rush of orders. 'Stand by Guns Crew, reverse the planes, open the Gun Hatch.' With that your lungs feel as if they're bursting, there's a roaring as the built up pressure releases through the Gun Hatch and also the Gun Layer with it onto the casing through a column of water pouring back into the boat. The rest of the Guns Crew followed and quicker than it takes to tell there's a sharp crack of the first round away. To the uninformed this evolution is both confusing and frightening but to the trained eye it is an extremely precise exercise, which was very effective during the last world war!!

Run Ashore

A Tiff mate's 'P' boat had just come alongside the base at Faslane after a three-month Fishplay exercise in the Gulf Stream area and the whole crew were raring for a good run ashore and piss-up.

The base at this time was very basic, with HMS *Maidstone* alongside the jetty and Boats alongside her with an LCT alongside jetty forrard of *Maidstone* used as an auxiliary store. The rest of the base was a compound surrounded by a chain link fence, the Guard House being part of the topworks of an old Merchantman, begged from the breakers yard just north of the base.

A sizeable section of the crew sent ashore en-masse led by the legendary Yorky Crossman, the boat's Chief Tiff, whose passion was horse racing, and whose intercom reports were always in racing terms. For instance, during routine servicing the OOW might enquire how the main engine repairs were progressing, the answer would be something like, 'Port's 'evy goin' 'n Starb'd's a non runner.' Anyway back to the plot, off they trooped to the lads canteen whose Manageress was Miss Crystal, but more of her later. All in Yorky's gang got well ratted and at closing time meandered their way slowly back to the Guard Room to pick up their Station Cards when Scratcher happened to see one of the lad's Pay Book on the desk, and in his pissed state, demanded it as he was the deputy Coxswain and he would deal with it. But the RPO in charge of the watch disagreed and an ugly argument ensued. The RPO's written report reads:

'Petty Officer Second Coxswain S. Wright O/Np8865372 was finally restrained by L/Patrolman Small by sitting on his head, whilst C.ERA Crossman attempted to bribe RPO P. Hall with "400 fags, 50 cigars, and as much rum as he could drink". Whilst all the time L/Seaman H. Tull climbed to the top of the fence shouting down "Leave him alone you Three Badge F— all shiny arm bastard" and many other profanities.'

The outcome of it all was that they all got various terms of stoppage of leave and other privileges, but as the Boat was away in a couple of days it didn't matter much. We all had a good laugh over the incident as there was no love lost between Boat Crews and the Regulating Branch and Crushers in general.

Run Ashore 2

The same Faslane base was the venue for another incident regarding a run ashore where I was lumbered with a shore patrol duty.

I was spare crew at the time and was waiting for my new draft to come alongside so that I could join her, but she was delayed which is why I got stuck with the duty shore patrol.

As ill luck would have it, a couple of Boats had, that same day, come alongside after a long and arduous patrol, so me and a couple of sailors were going to have our work cut out maintaining order and closing on time.

Five minutes before closing time, complete with boots, belts, gaiters and armbands I took my patrol into the lion's den so to speak, to be greeted by raucous cheer. I automatically checked the clock above the bar, whose glass was missing, and noticed to my horror that it was ten minutes slow! So when the cheering had subsided I shouted that I was sorry but the clock was wrong, grabbed a chair, stood on it and advanced the clock to the right time to the accompaniment of loud boos. As I stepped down and turned to face the onslaught the boos turned to cheers as some wag had, quick as a flash, jumped up and put the thing back to ten minutes slow again.

All I could do was smile and tool my patrol into the back private area to confront Miss Crystal, whose name belied her ample frame, and explained that the canteen would have to close ten minutes late. She was not well pleased, in fact she complained bitterly and told me she was going to report me. I said that it was her prerogative and that I would take full responsibility. In due course the Canteen closed late with very little trouble, and I was severely reprimanded.

The sequel to the story is that a similar situation occurred to some other poor sod but Miss Crystal insisted on closing dead on time and a riot ensued and the place was wrecked and then burnt to the ground, so she was out of a job and the troops had no Canteen.

Sod's law, eh!!

Claims To Fame

The post-tot-time discussion ensued, the subject of the day being individual claims to fame or possible infamy.

'I wasn't so much bored, more frustrated at the inactivity,' Nobby volunteered.

'I'd been struck in Pompey Barracks for what seemed like forever, day working down the Dockyard on my latest draft for a trip to the Far East. The refit was running behind schedule and all leave was cancelled and as a no badge f— all I could do nothing about it.

'The snag was that my best mate and towny's leave wasn't cancelled 'cos he was a different branch, and on that last weekend leave came back bragging about what a great time he'd had as usual, but in the conversation it slipped out that he'd seen my long term girlfriend with some smoothy or other, and I ended up in DQs.'

'What do you mean you ended up in DQs, why?' Pancho queried.

'I took an unofficial trip up homers and duffed up the opposition.'

'You mean you went AWOL and administered GBH?'

'Yes, but it worked.'

'What do you mean it worked?'

'I was banged up and missed the Far East draft and when I got out I married her and we're very happy!'

'Come on Pancho, what's your boring claim to fame?' Ali queried.

'Oh there's only one incident I can think of but it wasn't me who was bored, it was the rest of my idle crew. I was at the time Chief Diver of HMS *Reclaim*, the official RN Diving Vessel. We were at anchor, I forgot where, but somewhere it was somewhere nice and warm. We were supposed to be doing routine maintenance checks on all the diving gear. But the lads were bored and lounging about waiting for stand easy. At the time the main diving bell was deployed over the side and I was in it checking the equipment. This bell type had hatches top and bottom and could either be used as an enclosed vessel or open at the bottom to allow exit and entry at depth. The top hatch is used for crew entry which is a little difficult as the umbilical and a triple chain set are strung over it. At this particular time, unusually, both hatches were open when some clown accidentally (I hoped) slipped the hoist cable safety clip with the crane fall drum in free run.

'The chamber dropped like a stone, the oggin spouted in at a rate of knots, faster than up the outside because of the differential dimensions inside and out. The column of water shot me straight out of the top hatch, missing the triple chain, I don't know how, and managed to grab the coaming of the side entry to the diving flat, where the watertight door had been left open fortunately. I dragged myself into the diving flat, soaking wet and bloody furious, and hot footed it to the upper deck to find the whole of my diving crew awake and leaning over the rail, caps in hand shaking their heads until I bellowed at them and called them very naughty names. It's the closest call I've had for a while I can tell you.'

Boom

I ended up with a draft to HMS *Lochinvar*, South Queensferry, which is now a marina. At the time of my draft though, it was a Coastal and Inshore Mine Sweeper Base with a Bomb and Mine Disposal section and a very comprehensive Museum and Ordnance Display unit in a converted hanger. I clewed up with Mike again which was good and we ended up as a team, maintaining the Mine Sweepers. Anything and everything was done for them, engine changes, Pulse Generator engine overhauls, we even hauled them out up a slipway into a shed for the Chippies to check and repair hull damage or us Clankies to change props.

The haul out procedure was quite exciting as the winches were electric but the power supply to the Establishment was not adequate for the job. So large American Caterpillar Diesel Generators had been installed. The snag with these engines was starting them as the system employed was quite bizarre in that a two-stroke petrol engine was fitted to the main engine, driving a friction wheel onto the main engine flywheel. We think that they were built this way to be shipped out to Islands in the Pacific Ocean to allow the Americans to supply power to enable landing strips to be built and run where no power existed, to wage war on the Japanese in the Second World War.

The routine was that the Cat was checked over and then barred over to check that everything was free, then the two-stroke was flashed up with a starting handle and warmed up slowly, and when ready revved up to full chat on the governor, a long handled lever was pulled down against spring pressure and the friction wheel contacted the main flywheel. The two-stroke started to dip in revs, and the governor increased the engine's power, the friction wheel started smoking a bit and the big Cat started to turn over and eventually fired on a couple of pots, and then three and four and away she went and the two-stroke was let go and stopped,

The Greenies then took over and the Cat was put on line and the winch was put to work. Quite an evolution and we sometimes had quite an audience on start-up.

On one occasion I noticed a friendly face in the crowd of rubbernecks, so once we'd got things going I went over to say Hi. It was Andy Broomhead, Brushhead to his friends, so Mike and I organised to meet in the canteen that evening and we had a good natter. He'd qualified as a clearance diver, a crazy job making safe and removing under water explosives and demolition charges and disposing of them safely. He'd just got back from helping to clear the Suez Canal after the Suez Crisis, and was recovering from the stress and strain of that evolution. He had been drafted to *Lochinvar* as an instructor, but reckoned that he was getting a bit bored, but had added a few exhibits to the museum display.

One of the exhibits was a five thousand kilogram German bomb that had a section of casing cut away to show the detonator and the filling which everyone was assured had been emptied of explosive and refilled with harmless wax. This particular display unit was parked near the exit and everyone used to kick it or pick a bit of wax out of it. That is until Andy took more than a passing interest in it.

He took a sample of the wax and had it analysed, then suddenly he'd organised a gang which shipped it out onto the beach, hired a steam generator rigged it up and melted out the 'wax' and set light to it. It went up like a volcano because it was the real thing, the German equivalent to Amatol. Well done Andy, and how lucky can some people be.

Gotcha

RN Chefs generally are very good and most work under very difficult conditions, particularly in Boats, and most don't get the recognition they deserve. I've seen the Chef of *Oberon*, a conventional diesel electric, sifting through stinking black sludge, feeling for the remaining few edible whole spuds in the spud locker.

One problem is that due to temperature fluctuation and high humidity because of being dived a lot of the time fresh foodstuffs tend not to last very long in Boats.

One other problem is that once the fresh meat, vegetables and fruit have gone you're on tinned and frozen and the digestive system tends to seize up a bit, or a lot, depending on how your guts are working at the time, and as I've said earlier most chefs don't get the recognition they deserve.

A good mate of mine who had served as Chef on *Oberon* with me for a while when I was wreaking ended up in Pompey Barracks waiting for a draft to another Boat. He was a patient fella but was well teed off when he was detailed off to do galley duties instead of being able to take things easy. Worst of all he was down to do the early shift for Breakfast and hated it. He had to turn to at 0400 to get things ready for reveille having, apart from anything else, to boil at least twenty dozen eggs.

What really pissed him off was this one young lad who was always first and used to leg it down to the server before the bugle had finished, having not washed or shaved. So my mate used to wind him up a bit by blowing an egg and putting it in the top tray with the boiled ones, and as the mark dashed towards the serving hatch, he'd pick up the blown one and throw it at his head, and of course the mug ducked in panic and the empty shell hit the deck and bounced down the gangway. This routine was repeated a few times and eventually the mark got used to it and started to head the blown egg each morning and got quite proficient at it.

My mate eventually got his draft chit, so on his last morning of duty Chef, instead of a blown egg he threw a full raw one with spectacularly messy results with the comment, 'It serves you right you scruffy, scabby, gullible bastard!'

Embarrassed

A Wren friend of mine told me a job she had early in her Naval career which started out as a bit of an embarrassment, and how it was resolved.

It was when she was promoted to PO and was drafted to HMS *Raleigh* at Torpoint in Cornwall, where one of her duties was film projectionist. She was called on to show all sorts of instructional films which were not just instructional for the audience but also interesting for her. One of the films she had to show on a regular basis was to graphically illustrate a lecture to young new recruits on sexually transmitted diseases.

To start with it was very difficult and embarrassing and she used to sneak into the projection room by the back door, but eventually she got used to it, and in fact became quite blasé about it. The audience, however, was always new, and they started off as Jack the Lad, looking forward to a good laugh and lots of crude comments at the start of the film, but as it progressed things quietened down, then there were a few groans, some just looked away, others rushed outside to by physically ill.

Eventually she used to make a habit of walking the full length of the lecture hall to get to the projection room when this particular subject was to be shown, to a chorus of jeers, catcalls and whistles, and at the end of the lecture and film show, getting great satisfaction in walking back down the hall with a wry smile seeing all the pale faces, and all those who had been physically ill looking so pathetic.

HMS/M *OBERON*

Wrecking

Eventually passing part three you were allowed to keep an accompanied watch, for us Tiffs it was in the Engine Room, but it was a bit boring as this type of Boat being diesel electric where all the manoeuvring was done by the Greenies (electrical department). The only taxing bits being crash dive practice and starting and stopping snorting, so I made noises and got myself the position of second dickey outside ERA, commonly known as the outside Wrecker, as the incumbent, Wally Hart, was shortly leaving the service. It suited me down to the ground, being varied with unusual tasks and challenges, and I sure got to know the Boat very well indeed.

A couple of instances come to mind. I was sat in the mess off watch when Scratch (the Second Coxswain) came in asking if anybody had seen the sewing machine. Someone said he'd seen it in the spare stowage locker in the fore ends but he understood that it was bust so I decided to go with Scratch to see what was what. Sure enough we found it where were told it would be and sure enough it didn't work. But to cut a long story short, I found that someone had tried to use an oversize shuttle which was jammed in. I managed to prise it out with no damage, and loaded and threaded the right sized shuttle and tried it out successfully on Scratch's bit of sewing, after which I got to do most of the machine sewing, which to be honest I didn't mind as it kept me busy.

Another little challenge was the film projector. When a Boat went on an extended trip Their Lordships in their infinite wisdom considered that up to date entertainment would keep Jack happy and so supplied vessels with up to the minute film releases and of course a projector and portable screen. Unfortunately, just after we left for an extended exercise the incumbent projectionist, a Killick Greeny, discovered that the projector was defective and I was invited to try and fix it. I found that the take-up spool-drive pulley bearing bush had collapsed and its securing screw was missing. So I went back aft and down to the shaft space, where a little four-inch lathe was installed, and got to work. The measure of the importance of film shows can be measured by the fact that I was excused watch keeping to complete the repair. The next evening came the big trial with me doing the projecting. It worked perfectly and went down a bomb, half a dozen laced-together Tom and Jerrys, 'Good Old Fred.' I was as a result voted senior projectionist and projector maintenance engineer, well whoopee-f—in' do!!

S.V. West

Leading Chef West was the best thing that had happened to the Boat. The whole Ship's Company agreed that he was the best Chef in the Mob, including General Service.

The daftest thing was how he ever got into the Mob in the first place. He was semi-literate and as queer as a nine bob note, but didn't push his luck in that respect. He was also quite an artist in his own way and would write out a menu of a blackboard in beautiful copperplate print, but in naval slang. For instance:

BREAKFAST
Shit on a Raft (Sauté Kidneys on Fried Bread)
Cackle Berries (Boiled Eggs)
Elephant's Footprints (Spam Fritters)
Arrigones on a Shingle (Tomatoes on Toast)
Train Smash (Bacon and Arrigones (Tomatoes))

DINNER
Babies' Heads (Steak and Kidney Pudding)
Tiddy Oggy (Cornish Pasty)
Figgy or Plum Duff (Suet Pudding)
Duff and Thickers (Suet Pudding and Condensed Milk)

An explanation is required for Babies' Heads: Steak and Kidney Pudding used to come in tins as 'Submarine Comforts' and to get it out of the tin both ends are opened and the contents slid out as the clacker is water paste. The resultant lump is delicately cut in half and both halves placed on a baking tray, filling end down. This procedure is repeated until the tray is loosely filled, leaving room for expansion. The tray is then placed in a preheated oven to bake the clacker and heat the contents of the said clacker. When baking is complete the tray is removed from the over, the clacker has expanded and the corners rounded and for all the world it looks like a tray of babies' heads. And very tasty too, served with Smash, two tinned veg and RBG, HEAVEN!

But the best about West was his prowess with bread. All RN vessels used to carry premix bread in the deep freeze which was all ready to bake apart from defrosting and adding yeast and to be used when stocks of fresh bread ran out or went mouldy. Most times premix bread was pretty dire and fit only for fishing, not as bait but as sinkers, even Shitehawks turned their beaks up at it, but Westie's bread was divine. He could make Cottage Loaves, Breakfast Buns, Plaits, Farmhouse Loaves, Bloomers, Tea Cakes, Hot Cross Buns, Fruit Loaf, you name it, he could and did do it. We even bartered his bread for dodgy films with the Aussies. The secret of his success we concluded was the fact that when he left the final mixture to prove, he stripped off and covered it with his shitty vest!!

Water

In conventional boats the scarcest commodity is fresh water as the storage tanks are of limited capacity and the electric distillation units are about as handy as an ash tray on a motorbike, and most of what's available is used by the Chef.

It's a well known fact in boats circles that the ones with the cleanest jobs washed the most. I was panel watch keeper at the time of the incident which started with this poncy Ping Bosun stomping up and down the after passage with his little towel round his neck on his soap bag in his sticky little fist and every time he passed me he peered down into the troops bathroom (two wash hand basins) then chuntering to himself. Eventually he got on my nerves so much that I stopped him and asked him why he was wearing a groove in the deck and what the chuff was up. He pointed down the ladder to the bathroom entrance and said, 'He's been down there for more than half an hour, and if he doesn't shift soon I'm going down to shift him.'

I looked down and all I could see was this large lump of pink flesh in the entrance to the bathroom but nothing else, then I got called away to do a panel operation. After another ten minutes Ping gave a snort of rage and shot down the ladder to sort things out, then suddenly there were roars of laughter coming from below. It appears that the Coxswain had decided to treat the lads as it was his birthday and so had taken this monster turkey out of the deep freeze and put it in the wash hand basin to thaw out.

The outcome was that we all enjoyed our turkey dinner!!

Sunk

The Admiralty-issue 'Notice to Shipping' signals, some of which regard prohibited areas where submarine exercises are held and when. Such a notice was issued for a specific area off Portland Bill. The area was regularly used for boats discharge trials, as it was in this case.

Discharge trials involve testing torpedoes, tubes and crew in firing them. The boat in question was the purpose built target boat *Otter*, which was armoured to be able to withstand practice torpedo hits, and was skippered by a very forceful and ambitious individual. All was set to commence the exercise with the recovery vessel at the far end of the range. The boat closed up to action stations and dived to periscope depth ready for the first run when the OOW on the search periscope spotted a small vessel halfway down the centre of the range. In fact the vessel in question was a small motor fishing vessel skippered by an elderly local and crewed by his younger relations, none of which had read or were informed of the said Notice to Shipping.

The boat's skipper was informed of the intruder, and got the Sparks to instruct the recovery vessel to approach the offender, and get him to shift out of the way. This was duly done and the recovery vessel resumed its position, but the offender stayed where he was, in the way!

In desperation the boats Skipper decided to tell the old fool himself, and ordered the Wrecker to blow main ballast and stopped blowing when minimum positive buoyancy was achieved, opened up the conning tower hatches, climbed to the bridge and ordered course and high speed on main motors to close the MFV and using a loud hailer came close alongside and read the riot act to the offender and told him to sod off in no uncertain terms.

That done the Skipper ordered hard a starb'd group up, max revs, dive-dive-dive, gave the klaxon tit a good couple of bursts, dropped down the hole and shut the lid. Then in regaining the Control Room ordered periscope depth and course back to the start line where, having turned down range and seeing nothing assumed that the old duffer had cleared off, carried on with the trials.

Unbeknown to him, as he turned and dived boat, its arse broke surface because of the vicious manoeuvre and the port after hydroplane guard had stuffed a great gash in the side of the MFV at the waterline and was sinking fast so the Skipper transmitted a May Day call and the crew inflated and launched the emergency dingy.

Having dived, the boat's Skipper was unaware of the drama he had created, but the RAF Search and Rescue Station at Portland Bill had picked up the May Day call and dispatched their Rescue Launch to the scene.

The MFV Skipper, seeing the RAF Launch, let go a signal rocket which unfortunately went the wrong way and sank his dingy from under them. Never the less all of them were eventually rescued.

Their Lordships were not at all pleased when the local rag the next day ran the story with the Headline:

LOCAL HERO SUNK BY RN RESCUED BY RAF

And the Boat's Skipper was suitably dealt with.

Wrong Direction

'O' and 'P' Boats were the first RN Submarines to be diesel electric, which meant that there were no shaft clutches and all manoeuvring was done by Greenies in the motor control room, the good thing about this arrangement was that it was far more flexible, for instance one Engine and Generator Unit could supply power to both Propeller Motors while the other one was charging Batteries.

All Officers and Senior rates had to qualify on the motor control panel as a precaution in case of emergency.

Telegraph orders were transmitted from the control room to the electrical control panel for propeller movements. As a precaution, a large red warning light was fitted to make the operator aware that if he ran the motors in the wrong direction it would illuminate. The big red light had a label above it which read, unsurprisingly, in big bold letters WRONG DIRECTION, but some Greenie wag had in a quiet, bored moment removed the red lenses and made little card cut outs and placed them behind the lenses and replaced them.

Unfortunately, some say, the next crew member to re-qualify was the Captain and, guess what, he selected the wrong direction, and the big red lights came on and told the Skipper that he had selected the wrong direction with a very very naughty word of the female anatomy and he was not amused, in fact he was extremely dischuffed and had almost permanent humour failure.

Manchester

Patience was what was needed when we took *Oberon* to Manchester. Yes Manchester, up the Ship Canal, which caused quite a stir with the populace along its length observing this big, sinister black beast silently sliding past. Unfortunately it was midsummer with glorious sunshine and nil rainfall for some considerable time which accounted for no change in the canal's contents. As we made our way up the black liquid and broke the surface with our bow wave the smell of hydrogen sulphide was appalling. We had just completed a lengthy exercise in tropical waters and the hull was badly fouled with weed and barnacles, but halfway to Manchester in the acidic canal left our bottom bare metal, even the anti fouling paint had been stripped off.

Having arrived at the docks and tied up, a tremendous crowd had assembled waiting to come aboard and have a look round, and were upset to find that visiting wasn't till the next day, as we all had to turn to and make everything clean and shiny with all our personal gear stowed away. Then the next day all but a skeleton crew has to be sent ashore as there was just the one passageway so only guides were allowed to stay aboard.

It appears that the record for the largest number of visitors passing through a conventional boat was well and truly broken on this show the flag trip. Quite a few stories of large people getting stuck in hatches was prevalent, and lots of complaints about the smell and the clutter of pipes and valves, but the general consensus was that it was an informative and good day out.

Manchester 2

The downside to the trip was that accommodation for the lads and senior rates should have been arranged ashore, but due to an administrative error (cock up) had been overlooked. Senior rates arranged B&B or similar, but Jack ended up spending the early nights getting well pissed and then getting their heads down upstairs at the back of all night buses, but the bus crews didn't seem to mind too much. In those circumstances the lads had to have a lot of patience and sense of humour, and most of us were glad when we made our farewells and got back to sea and normal.

Attraction

One of the most inept chasers of the opposite sex I've ever come across was POME (Stoker PO) Robert 'Shiner' Wright, who at Tot Time in the Spare Crew Mess of *Adamant* finally admitted his shortcomings and related the ultimate in humiliation.

Poor old Shiner wasn't a bad looking lad, a bit of a loner, good at his job and always a neat and natty never gaudy dresser who was always on the lookout for, and very unlucky with, female company. His boat, *Seraph*, one of the last of the little 'S' boats, had just returned to Faslane after a long and arduous exercise in the North Sea.

He took ages tarting himself up for a run ashore in Helensburgh, and after a quick snifter from his bottled neaters (strictly illegal), sallied forth. He clewed up the only customer in the bar of a very exclusive Hotel trying to drown his poor progressive sorrows. Then when he was at this very lowest ebb, a very attractive lady swept in and parked herself most elegantly at the other end of the bar, ordered and started to sip a drink, looked round to survey her surroundings and smiled sweetly at Shiner whose heart rate and expectation soared. He smiled back and started planning his next move, possible offer her a drink, or wander across nonchalantly and introduce himself. Not being able to make up his mind, she ordered another drink and gave him another smile which he responded to with fervour, and was about to make his move. But suddenly she picked up her drink and elegantly strolled towards him and parked herself on the bar stool next to me. His heart was pounding so much it felt as if it was going to crush the fags in his inside breast pocket. She turned to him, smiled sweetly again, and said in a lovely low sultry voice, 'Hello, are you in Submarines?'

Quick as a flash, not believing his great good fortunes he replied: 'Yes, I am, how did you guess?'

'Oh it was no guess, my husband is the First Lieutenant on *Oracle*, I could smell you!!!'

Discharge Trials

To test the efficiency of the forends crew and their equipment, discharge trials were carried out by all boats on work-up. Our turn came in our new 'O' boat that I had just joined. The trials were held in Loch Long which as its name implies is long and thin. Working with us as always was a civilian MFV (motor fishing vessel) that positioned itself at the far end of the range. The idea was that at the end of its run the torpedo activated a blowing head which caused it to bob up vertically to show its red luminous nose to make it easy for the MFV crew to recover it. The reason for this little story is not for the mechanics of the trials but for the perks, because the MFV crew were full time fishermen and part time RN employees and after a stint of torpedo recovery always came alongside with a box full of assorted seafood in exchange for a bottle of scotch from the wardroom. That being so the Wardroom steward always had the pick of the box and the remainder was donated to us plebs, but we of the lower deck had a secret weapon up our sleeves in the form of Killick Stoker Frederick 'Soapy' Watson, who in his earlier life had been a very good Fishmonger, who took what the steward considered to be rubbish and knocked up some incredible meals for us in collaboration with our chef Shitty Vest West, such as Bouillabaisse, Kalamari, Pizza Di Mare and Paella.

One of the funniest stories about discharge trials I've heard though, was the one about the Mk21 that suffered a gyro malfunction, the bit that controls direction. Loch Long is not only long, it is also very beautiful, which is why whiskey magnates and the like build mansions with billiard table smooth lawns on its banks. The Mk 21 is driven at high speed by an internal combustion engine fuelled by shale oil and compressed air with contra-rotating props. This particular wayward tin fish decided to head for a particularly nice lawn which just happened to come right down to the water's edge, and proceeded to plough its way towards the big house, coming to rest within feet of the conservatory.

The really embarrassing bit was that Jolly Jack had to take a gang and a Wagon round to the Big House, knock on the front door and say, 'Please may we have our torpedo back!'

It's rumoured that the Admiralty had to pay out thousands of pounds in compensation and insurance claims, and the lawn was never quite the same again.

Tail End Charlie

This is the experience one of our Stokers, who told me of his ordeal when he was serving on his first 'P' boat. This incident happened when his boat was doing discharge trials on Loch Long on the west coast of Scotland. Discharge trials were exercised to test the efficiency of the fore ends crew and firing and reloading torpedoes.

Loch Long as its name implies is long and straight, ideal for launching torpedoes which are fitted with blowing heads instead of explosives so that they surface when they run out of fuel and are recovered by civilian MFVs seconded to the RN

Anyway it appears that halfway through the trials the hydraulic loading system became unserviceable and the old fashioned block and tackle method had to be resorted to, so off watch stockers were 'volunteered' to make up the numbers including our green storyteller. Being new to the job he was tail end Charlie on the haul line so he was nearest to the short ladder leading to the watertight bulkhead door.

Things were going quite well although understandably, a little slower than normal so orders were given to speed things up. Unfortunately on the very next reload the Fish stuck halfway up the spout so an extra effort was called for and the gang gave an extra special 2-6 heave and the fish slid forward at a rate of knots, overshot the top stop (which was supposed to lock it in position) and hit the outer door with a loud clang springing it off its watertight seat momentarily. The result was the sound of rushing water and a gout of oggin shot from round the fish and out the rear door.

The next thing our Stoker knew was that he was laying face down at the bottom of the ladder with the bulkhead door being slammed shut and clipped with pain and footprints in the middle of his back. Being green our poor unfortunate didn't realise the implications of the overshoot, but the old hands knew the possible scenario and took immediate evacuation action and, because he wasn't fast enough, used him as a stepping stone.

The moral of the story is, know when and how to take evasive action.

Let's Twist

We were involved in an exercise with the American Atlantic fleet with all their up to the minute vessels and equipment.

We were operating at the northwest coast of Ireland, and to get to the exercise area well off the continental shelf we had to leave Londonderry the previous night.

Our job, in our diesel electric 'P' boat, was to avoid the Destroyer screen and attack the capital ship they were protecting, which in this case was their latest most up to date Aircraft Carrier. Easier said than done because apart from the Destroyers having modern sonar, the Carrier's armament included helicopters equipped with 'dunking' sonar. The idea of dunking sonar was that the helicopters were directed to possible contact areas by the Destroyers and having arrived come to the hover and winch down the sonar sensor and dunk it in the oggin where a specialist crew member checks for contacts and directs Destroyers on depth charge runs accordingly. The helicopters also adopted team tactics to check larger areas.

What our Skipper did to outwit the opposition was to work out the mean course of the Flotilla as it was zig-zagging and circle round in front of its mean track. When we had achieved this he ordered silent routine and all machinery was stopped, carpets laid on steel decks back aft, fluorescent lights switched off (they buzz) and all crew off watch turn in. The only thing left running was the hover pump, a small silent pump adding or discharging ballast water to maintain neutral buoyancy and depth.

Our passive hydrophones were used to track the fleet's progress and eventually detected that the Van has passed over us and we had avoided detection. The Skipper allowed a short while to elapse and then ordered slow ahead, periscope depth, and we crept slowly up, heading for the Carrier.

I was Wrecking at the time and had just routinely desiccated the attack periscope and our REM Bogey Knight, who was a photography nut, had recently serviced the periscope camera which was lucky for us because the Captain ordered the camera to be loaded with film and attached to the attack periscope.

After what seemed like an age and a few quick shufties and a few all-round sweeps through the attack scope the Skipper informed the crew over the intercom that the Capital Ship (Carrier) was about to run over us. He then ordered one hundred foot depth, and before long we heard faint prop noise from the target. We went back to silent routine and the Gaffer called Scratcher to the control room and gave him specific instructions. We found out afterwards that, as a music buff, Scratch was ordered to rig his record player up close to the underwater telephone and sort out a record of Bill Haley and the Comets. Then at the precise time that the Carrier was overhead the Skipper ordered the underwater telephone switched on and turned up to full volume and 'Let's Twist Again' shattered the depths. It was switched off after the first two lines and we went back to silent routine and the sonar shack reported that all the escorts had turned in towards the Carrier.

The object of the exercise was for the Destroyers to try to locate us and if successful to drop hand grenades on us and when we considered we were hit we were to fire a smoke candle, but nothing came anywhere near close to warrant a candle although they must have wasted hundreds of grenades. So after a while the Gaffer took us up slowly and silently to periscope depth, had a quick all round sweep with the attack scope and proceeded to take photographs and used the whole roll of film up.

We went deep and the camera was given to the REA with instructions to process the film and print anything that looked interesting. What Bogey did was truly magnificent, the whole roll of film was printed and some of the more interesting shots he enlarged. I managed to thieve a couple of small prints but most of them were a lot better quality.

There were lots of shots of the Carrier, lots more of the escort Destroyers, but the best ones were of the helicopters. In some of the shots you could see that the pilot was Joe Bloggs and his chopper's pennant number crystal clear, and you could see the electric string dangling down to the dunking sonar.

After any naval exercise there is always what they call a washup where all participating Skippers and their experts gather with charts, tracks and logs etceteras to analyse the outcome of the exercise. The Senior Officer, in this case the American Admiral of the Carrier, is always the Chairman in charge of the washup. After lengthy discussions and pourings over all the evidence presented the Admiral concluded that the Americans had won and that as he so succinctly put it:

'There were no British Submarines in my Carrier's area.'

Whereupon our Skipper reportedly opened his briefcase, dug out a thick handful of very large photographs and flung them towards the Admiral who, on inspecting them, went a funny shade of puce, looked as if he was about to explode, and stormed out of the room with all his hangers on scurrying along behind him. We all reckoned that Joe Bloggs was in for a bit of an upping, and the Skipper of Destroyer 824 and the jockey of helicopter number 59 were not going to get away scot free!!

Across The Pond

We started out from Gare Loch to surface transit the Pond. But as soon as we stuck our nose out after Ireland we hit roughers and I grabbed my bucket to contain the arisings from my mal-de-mer. Problems arose almost immediately because we were the first Boat to be fitted with a fibreglass casing and it had started to break up. We kept going with the OOW keeping a careful eye on the damage's progress. It became critical when the after escape buoy cover started to disappear and the aerial deployed and started transmitting our Subsunk distress call. The Skipper was informed and went up top to look for himself having told the radio shack to transmit a cancellation of the distress signal and then ordered a couple of dabtoes to get rigged in safety gear and harnesses and go and sort things out.

We then hove to and as the lads got onto the casing and clipped on, the rest of the restraining covers flew off and the buoy popped out and over the side still transmitting. It was eventually recovered and silenced, but the casing problem persisted so the Skipper, to limit the damage, ordered a surface snort transit. This involved the induction mast being raised for engine combustion air but with the conning tower hatch left open and normal watch keepers on the fin with the boat trimmed down to cut through the waves rather than ride them. It helped to reduce the casing damage but the poor sods on the fin kept taking green ones and the control room didn't fare much better as columns of oggin kept shooting down into it, the poor sod on the helm taking the worst of the soaking.

Eventually the weather abated and the extent of the damaged casing was assessed and a signal sent to Their Lordships requesting replacement sections which were waiting for us when we eventually arrived in Norfolk, Virginia. We fitted it all with help from our Buddy Boat crew in sweltering weather.

The American Buddy Boat system was great and I hope that it still exists because it made us visitors to the States feel very welcome.

The idea was that your Boat was moored alongside a USN Boat, or nearby, and their crew looked after yours, and they did it very well too, being very generous, pandering to our every whim, and quite often taking you up homers.

They also organised group activities such as barbecues and baseball matches. All in all we had a wonderful time after a lousy start.

We were also invited to take a Cook's tour of the Base and Dockyard, which was incredible because of the facilities that were available. The crews and civilian employees wanted for nothing, they worked hard it's true, but off watch the amenities were second to none.

Every conceivable facility was available, and their PX put our equivalent NAAFI to shame, it was like an upmarket superstore, in fact the whole base was run like a small city.

The Magic Torch

I was on my first 'P' boat and was wrecker at the time and the incident involved snorting which is the process of running the main diesel engines while submerged at periscope depth by drawing air down an extendible mast called the snort induction. The reason for doing this is to enable the battery capacity to be conserved or even charged up whilst avoiding radar detection.

This particular time we were running deep and the Skipper ordered the OOW to initiate snorting. So Jimmy, who was OOW picked up the waterproof rubber covered microphone from its stowage between the two planesmen and ordered over the tannoy:

'Stand by to snort, go to black lighting.'

Black lighting was in fact pin points of red light in the control room used to acclimatise the Skipper at the periscope to the darkness up top. The rest of the boat stayed with normal lighting. Jimmy then told the planesmen to come up slowly to periscope depth and on the way up had Asdics do an all round sweep to check the area clear. On gaining periscope depth the Skipper did an all round look with the search scope to further check the area was clear to continue with the snort exercise. At this time it was the change of watch and the new Subby took over from the Jimmy.

The normal position for the microphone was laid across the stowage for convenience sake, but just before black lighting Jimmy had been looking for a small suspected water leak behind the main panel and used a waterproof rubber torch to look for it, but the batteries failed, so at black lighting he put the microphone in its stowage and the duff torch on top with the intention of changing the batteries and renewing the search as soon as convenient.

The radar mast was raised and the screen checked clear so the Skipper told Subby to start snorting, so Subby picked up what he thought was the microphone to transmit all the complex instructions required to perform this evolution, but was in fact the duff torch which I spotted. So I grabbed the Chef, whose galley was aft and between me and the Engine Room where the ballast pump was also sited. I quickly explained the situation and he set up a command chain to the Donk Shop Horse and Ballast Pump Operator so that all instructions supposedly over the Tannoy were transmitted verbally and all responses to the orders came back to the control room via the Tannoy system as normal.

After a reasonable length of time snorting successfully with good depth control and trim stability the Skipper ordered the Subby to secure from snorting, which required more complex orders, go deep and revert to white lighting.

Eventually the lights came on, the Subby went to put the microphone back in its stowage and realised it was a torch. Looking absolutely horrified, he had a quick all round sweep to see if anyone else had noticed. When he thought nobody had, as everybody involved in the scam had melted away, he surreptitiously replaced the torch on top of the storage and carried on as if nothing unusual had happened and to this day I'm sure he's still confused!!

Heath Robinson

When I was serving on my first 'P' Boat on Wrecker I used to get constant earache from the Engineer about the leaking torpedo loading hatch. I'd tried everything to cure it and it was always OK at periscope depth, but below that the deeper we went the worse it got. It never really got that bad, it was just a bloody nuisance.

The basic problem was that the loading hatch, as in all this type of Boat, was set at an angle to the pressure hull to enable fish to be loaded in one long piece. Unfortunately this means that although the hatch itself is round the hole in the pressure hull is elliptical which is a weakness due to compression at depth. To combat this, when the hatch is battened down, strongbacks (large diameter tight fitting bars) are slotted into crotches across the aperture to prevent hull distortion due to pressure at depth.

It was a nuisance because as you went through the fore ends bulkhead access hatch to go forrard up the centres, unless you were aware and very careful, you got a neck full of oggin. One Skippers rounds, the Captain got the dreaded neck full and rounded on the Killick fore end man and instructed him to 'do something about it' so he did. On a long boring three month Fishplay exercise said Killick started by diverting the steady trickle to one side using a suspended piece of lagging sheathing (half a two-inch diameter tube of thin aluminium). From then on, every time the Skipper did his rounds another bit had been added with considerable ingenuity, and it ended up like a Heath Robinson sculpture with water wheels, tipping scales, little animated animals and roundabouts, seesaws and assorted other gizmos, until finally disappearing into the bilge to be returned from whence it came, to the ogwash. The Skipper couldn't in all fairness complain because his last order had been obeyed!!!

The Trim

A mate of mine, who was Wrecker on a 'P' boat, told me of the new Jimmy that had joined and immediately got everybody's back up.

Officers aren't meant to be popular or liked, but in Boats a certain tolerance and benevolence is expected. The new Jimmy though, started cracking eggs with a big stick, nothing suited and he did a lot of nit picking, which pissed everybody off. His one Achilles Heel was that he was absolutely hopeless at catching a trim and keeping it.

Catching a trim means getting the overall weight of the Boat right so that when dived it is not only neutrally buoyant but also that it is balanced fore and aft, and it is the job of the First Lieutenant to catch the initial trim on the first dive.

He wouldn't take advice from anyone regarding trimming and at times it was all hell and no notion, with the ballast pump operator in the Engine Room leaping about like a one-armed paper hanger, cursing all the time with the intercom running red hot, and the planesmen sweating cobs trying to maintain angle and depth, the Boat all the time proposing and assuming odd fore and aft angles.

His comeuppance came in the shape of circus act in the guise of a tightrope act with a cyclist on it. Some wag, just before said Jimmy was due on his diving watch as OOW, had rigged piano wire tightly between the two periscope masts and put a little balanced bicycle with grooved wheels on it, with a rider whose feet were attached to the pedals and articulated legs.

Immediately after being handed over the watch, without noticing the little bike behind him, he started re-trimming with the inevitable result of lost trim, and the cyclist started pedalling back and forth between the two masts.

The buzz went round like wildfire and quite a crowd started gathering quietly in the control room, the stokers from back aft and the dabtoes from forrard, all watching the little fella zooming back and forth ever more urgently until his little legs were going like pistons accelerating to a blur in mid-traverse. Some of the lads couldn't contain themselves any longer and started with a little giggle that very quickly escalated to roars of uncontrollable laughter which alerted Jimmy to the situation at the same time as the Skipper came out of his cabin to see what all the commotion was about.

Number One was never the same again and was replaced shortly after the incident, but he stayed in Boats and, having calmed down, eventually made a very good Skipper.

Coqc

A friend of mine, on a run ashore in Helensburgh, told us about the time he was torpedo officer on a 'P' boat doing a COQC (Commanding Officers Qualifying Course).

Prospective Boat Skippers, having done all the theory in the classroom and practice in the attack teacher have then to go aboard in a group and take turns at executing for real attacks, the only thing missing being live torpedoes. The bloke in charge of the course was a very senior Boats Skipper, and a right bastard to the students.

The crew on a COQC Boat hate it because everybody's at attack stations all day for days on end, and everybody gets very on edge, especially the planesmen and helmsman who are moved about like a fiddler's elbow. At the end of the course the students are told there and then whether they've passed or failed and, depending on the results, everybody goes ashore in Rothsey to either celebrate or drown their sorrows, and I was invited to join in along with a couple of others out of the wardroom. We all got well smashed but some rotten sod had taken a large box of dhoby dust with them and stuck it all in the local fountain. You've never seen so many suds flowing down the main drag, and the local Constabulary were less than amused and the culprit didn't own up so we all got a right bollocking, but it was very spectacular just the same.

I Had To

One tot time the conversation in the mess turned to why each other had joined the Mob. Some rather strange reasons were given, but the most bizarre was that of the RO who at first had been quite reluctant to disclose his secret, but eventually he succumbed to the relentless harassment and admitted that he had been forced to join. When virtually interrogated as to the reason he reluctantly explained that he'd been a naughty boy and was brought before the beak who gave him the choice of joining the armed forces or going to jail! This was greeted with a modicum of scepticism and the demand for the reason. Eventually, due to popular demand he explained.

'Well if you must know it was because of the first job I got at this builders yard on the outskirts of Pompey. Me Dad got me the job 'cos a mate of his knew the Foreman at the yard. All I did all day was sweep up and generally prat about until one day one of the gaffers dashed into the yard when I was the only one around. He grabbed me and asked if I could drive, I said yes, so he told me to take a diesel driven Road Roller to a building site down by the Dockyard ASAP. The only problem was that I had forgotten to tell him that I was only fifteen years old and didn't have a driving licence but he had dashed off and it was too late.

'For those who don't know a diesel roller doesn't have any brakes and just a forward and reverse crash gear box. I managed to start it OK selected forward gear, got it out onto the road and put my clog down. As I've already explained the yard was on the outskirts of Pompey but it was also at the top of Portsdown Hill, which is not particularly steep but extremely long, and I had to go down it, and even with the pedal to the metal it was really slow going. That's when I made my big mistake, I took the dam thing out of gear and it picked up speed very nicely. Then things got a bit out of hand, and I was going a bit too fast so I tried and failed to put it back into gear because it was a crash gearbox, not synchromesh like a car. It just kept going faster and faster until I started having to pass cars, the drivers of which looked more than a bit surprised, to say the least. I eventually came to rest at the bottom of the hill feeling very relieved with the Police following very carefully behind. I was charged with dangerous driving, and speeding in a road roller whilst under age and not insured.

'The Beak gave me the choice so I signed on!!!'

Look Before You Shoot

Poole Harbour is the setting, a boat alongside, ready stored for the next trip. All the routine maintenance was complete apart from the air shots.

Torpedoes are launched using a slug of compressed air in conventional Boats, (water ram on Nuclears) and to prevent detection the firing air is returned inboard, as the fish leaves the tube, through the AIV (automatic inboard vent). It's all quite complicated really and accuracy of set up is critical, and as most of the gear involved is totally or partially immersed in salt water, whilst in harbour, the systems are tested with 'Air Shots'.

Air Shots, instead of firing torpedoes, shoot out just a tube full of water which has the same effect with the system. The one thing the TGM (Torpedo Gunner's Mate) must do is to check that the tube he is about to test is empty! To this end the tube rear door has a label holder to indicate the contents, but the rear door should still be opened to confirm that the label is correct just in case.

The poor unfortunate who was about to perform such a routine air shot must have had a heavy night or something similar, and only firing on three cylinders. No label was in the holder of the tube he was about to cycle so he just assumed that it was empty.

The boat was about to depart for sunnier climes and had just completed taking stores aboard for the duration, but the TGM had other things on his mind and flooded the tube, opened the bow cap, pushed the button and let her rip, and everything worked perfectly. So he couldn't understand why the Cox'n barrelled down the forends and started giving him an ear bashing and reading his horoscope.

'What the hell's up?' the poor unfortunate pleaded.

'Come with me,' and literally frog marched him onto the casing where the rest of the crew were already assembled.

'LOOK,' and spread his arms wide.

All you could see was EGGS, no oggin, just eggs.

'You have just discharged the whole of the next trip's egg supply into Poole Harbour, I don't think you're going to be very popular apart from with the Shitehawkes.'

By way of explanation, eggs keep for quite a while if stored in a cool place. They are also very popular with the lads who call them Cackle Berries. They are also very bulky to store which is why, with the Wardroom's permission, the Cox'n ordered the Chef to have the said tube loaded with eggs, making sure that the boxes were tied together so that they could be retrieved when required.

Ten boxes x two compartments / box x seven trays / compartment x eighty one eggs / tray = 11,340 eggs which float in salt water. What a picture. And what a mess!!

The Overflowing Freezer

We were involved in an exercise in the North Atlantic. We were the very boring target for the NATO surface fleet. Eventually we arrived at the allotted area, the exercise started and, most unusually, we were almost immediately detected as signalled by a hand grenade. Starting another run we were again almost immediately detected which was a bit more than coincidence and suspicions were aroused. Surfacing to do a fast transit to the new start position the OOW called the Captain to the bridge and pointed out an oil slick in our wake, so the Skipper ordered a 360 degree loop round and as we regained our original course the small of diesel fuel was overpowering.

That's when I was called to the Bridge as I was the Wrecker and I was joined by the Chief Tiff to try and establish where the leak was coming from. The only answer I could come up with was the cooling water discharge from the Refrigeration units which controlled the temperatures of the air conditioning, fridge and deep freeze. The Chief couldn't be sure so we went below and consulted the drawings which confirmed my initial diagnosis that the discharge pipeline ran through an external (to the pressure hull) fuel tank. To ensure that this really was the case the Chief and I suggested that as the sea surface was calm the Boat hove too to see where the fuel rose from. The Skipper agreed and called down all stop and as we lost way and came to a stop, sure enough a lot of fuel globules welled up from the location we had predicted and quickly spread into a large slick. No wonder we were being sussed out so fast, all they had to do was follow the oil slick, cheating gits! The Skipper sent signals off reporting our condition and we were recalled to Faslane and good old AFD 58 to effect repairs.

So we headed for home and the Skipper got Jimmy to check with Scratch (Second Cox'n) the paperwork regarding the contents of the fridge and deep freeze, the reason being that the Cox'n, Clancy McGrail the grocer, was on compassionate leave. In the meantime the Chief Tiff got the Donk Shop Horse to draw fuel only from the defective tank to try and limit the losses and pollution. According to Clancy's books there wasn't much grub in either the Fridge or Freezer, so with Shity Vest West the Chef in tow with Scratch, the Jimmy went to see what could be consumed before we got to Faslane as anything left would have to be de-stored and humped ashore. The Fridge yielded just a bit more than was listed, but as they opened the deep freeze it fell out at them. It was ram jam full of meat of all descriptions, unbelievable!

Needless to say that on the way back all you could get to eat was Meat. If chicken was on the menu that's what you got, a chicken, a whole chicken on its own. Breakfast was Bacon, Sausages, Kidney, Steak, and Chops, Lamb and Pork. By the time we arrived in Gareloch the whole of the ship's company was sick of the sight of meat but it avoided questions being asked inboard although Clancy got a mild bollocking for not keeping true records. We all agreed though that he was a good Grocer who fed us a lot better than most other Boats.

Fishplay

Fishplay was an exercise with NATO, with us swanning about the Gulf Stream for three months sweating cobs where everybody, including the Skipper, wore rag sack specials. You dived into the ready use rag sack and dug out the largest piece of cloth you could find and used it as a sarong, nothing else, just a lump of rag wrapped round your waist, nice and cool for the nether regions!

Anyway, to ease the boredom, one of the forend men decided to produce a newspaper. He used large sheets of chart correction paper and hand printed all the articles and even drew photographs to go with some of the items. One of the items I remember was Uncle Clancy's Kiddies Corner (Clancy McGrail was the Cox'n) with a junior crossword, games, cartoons and even a picture to colour in.

The general pattern was spreading buzzes and extracting the urine from all and sundry, all very cleverly done, it must have taken him hours to compile, let alone think up.

One article in the 'Letters to the Editor' section which I recall vividly was:

To the Captain,
H.M.S/m/ Oberon.
From the Headmistress,
Eavesham College for Young Ladies.

Dear Sir,

Thank you ever so much for showing our Senior Girls round your converted Sewer Pipe. Please may we have them back now?

I believe the Skipper actually collected them up and saved them as souvenirs of a very happy Commission.

Hanging By A Thread

I was a Wrecker on HMSm *Oberon* and we were at Harbour Stations about to take a load of visitors on a PR trip. It was just a day trip, a pain in the arse for us crew but necessary I suppose, but one good bit was that one watch was left inboard so that a lot of rubberneckers could be embarked. This arrangement was because Boats only carry the bare minimum of escape equipment, and the visitors if necessary use the now spare equipment.

Anytime you take visitors on a sea trip they always seem to congregate in the Control Room which is the last place you want them when you're closed up ready for leaving harbour, but all the usual checks have to be carried out regardless. One important part of the checks for the Wrecker was to ensure the correct operation of all the masts. The only exposed masts were the periscopes, for obvious reasons, and were extremely complex.

When lowered the long ten-inch diameter tube is housed in a well, and when raised by hydraulics the Ocular Box, the bit at the bottom end that is looked through, can be adjusted for height by the Tiff on the panel, but for maintenance and other purposes it can be locked in the fully raised position by inserting drop nosed pins through hooked brackets on the crosshead of the ocular box and fixed eyelets in the Deckhead. I had checked al the masts OK apart from the Attack Periscope which was unsighted as it was well forrard and surrounded by bodies. I shouted at the top of my voice as loud as possible above the general hubbub, 'Stand clear of the forrard periscope.'

I then operated the control vale and heard the satisfying hiss of hydraulics and saw the scope rising, then suddenly a strangled scream rent the air. Looking in the direction of the cry I saw this civilian's head and shoulders above the rest leaning forward at an impossible angle jammed up against the deckhead.

What had happened was that he had been standing too close with his back to the mast and as the crosshead had come clear of the well the pin hook had slid up inside the back of his jacket and hoisted him to that position.

Being well aware of what had happened I began immediately to lower the poor unfortunate but in doing so, as he was very much leaning forward, proceeded to stuff him down the mast well, so another strangled scream emanated from the poor sod prompting me to stop so that he could be unhooked.

After all that he was non the worse for wear but he kept well clear of the periscopes.

Chefs

'Don't talk to me about Chefs,' my mate recalled, 'all they've done for me was give me grief.' The final ignominy being the bunging up of the Pigs' (Officers') Head when I was conned into training as wrecker on the 'P' Boat I had just joined. P and O boats' heads are luxury compared to the earlier A, T and S classes, in that instead of having to blow your own residues directly to sea, and run the risk of getting your own back if you get the method wrong, all such arisings are collected in an internal tank and discharged in bulk when full and/or convenient, by blowing out using compressed air. The four heads were arranged in line just aft of the diving panel, three basic traps for the troops and sheer luxury for the Officers with not just a bog but a wash hand basin as well. Entrance to said Officers' Head was via bat wing doors situated opposite the galley and just forrard of the Engine Room bulkhead.

The problem was, with this layout, and always will be that it's too easy and convenient for the Chef to empty his slop bucket down the nearest bog, and that was of course the Officers' Head. The second problem was that the Chef very often left either a cleaning cloth or eating irons of both in the bottom of the bucket and worked the flush handle before he realised what had gone done the hole. The third problem was the fact that the foreign objects never got to the tank but caused a blockage in the pipework, which caused a backup of subsequently discharged material from the three other upstream Troops traps as it was a common line with the Officers' trap the last to join it, and therefore the one to concentrate on for clearance purposes.

The clearance procedure, I was told, was tried and tested and worked well. The plan was to turn off the flush water, which ran when the pan flap valve was operated to flush the arisings. The flap handle is then jammed down so that pan was open to the tank with the blockage between the two. The next stage was to cover the top of the pan with a gash bag, a linen bag treated with a coating that made it slowly disintegrate in salt water, and then put the lid down on it so secure it, then a slight pressure was applied to the sewage tank from the discharge blow station situated by the Engine Room door, and the tank top inlet valve opened to blow back up the soil pipe. Usually this is sufficient to cause a slight eruption under the gash bag and the foreign objects removed from the bowl and the culprit confronted with it.

Unfortunately for me this did not happen, so I gave it another bigger burst and still nothing, so I went to ask the Wrecker what to do next and all he said was give it an extra burst, so I went back and gave it another good burst, the over pressure relief started to feather and still nothing.

That's when I made my big mistake, I peered through the gap in the bat wing doors to find that nothing was happening, so I poked my head through the doors to find if I could hear anything, then with no warning the damned thing just erupted. Like a fool I just tried to pull my head back but the doors were spring loaded and they jammed behind my ears. I got a six inch wide stripe of shit all the way down me and two lovely black eyes, which everybody took the piss out of me about. Talk about getting your own back, I got everybody else's as well.

I'm OK

This is the saga of two good mates who suffered a bit of a hiccup in their friendship. The Engine Room Tiff had been off sick and had just returned back aboard. His mate the Radio Electrical Tiff was replacing a defective heating element in the Snort Head Valve. This valve, on top of the induction mast, allowed air to be drawn into the Boat while at periscope depth to enable the main engines to be run.

The heating elements were necessary to prevent the valve freezing up in sub zero conditions as the valve would have to close if the head dipped under water to prevent the Boat flooding.

On this type of boat the snorthead valve was streamlined and flush with the fin top when lowered. Therefore to be able to gain access to the valve internals it had to be raised, but not much, just sufficient, a couple of foot, but it meant that to do the job and be comfortable the REA had to sit on top of the fin with his legs dangling in the hole left by the top of the valve. His big mistake was not putting a 'Do Not Operate' notice on the operating control valve of the snort mast.

The ERA, who was Wrecker, and therefore IC the control panel, which included the mast raise and lower controls, swanning back from sick, passing the panel saw that the Snort Mast was not fully down according to the indicator and so decided to lower it. A loud shriek emanated from the conning tower hatches and he immediately let go on the control lever, dashed up the Tower and saw his mate jammed in the hole by the streamlined cover on his legs. Despite the REA's protestations that he was OK his mate shot back down to the Control Room and belled the Base Ambulance Service and explained that he had broken his best mates legs and was stuck at the top of the fin. With a loud siren and a screech of brakes the Ambulance arrived in record time, and the Tiff immediately took charge of the crew much to their annoyance, untrapped his mate and saw him loaded into a Neil Robertson stretcher while he complained all the time that he was OK and was told in no uncertain terms by his increasingly agitated oppo to shut up and let the Medics do their stuff.

Lashed in the stretcher, which was specially designed for this type of evolution, the only exposed bits being his hands and feet. He was lowered down the first hatch successfully.

The snag came at the lower hatch where he started to complain bitterly about pain and was again told in no uncertain terms by his mate to stop complaining and not be a wimp and stomped on his shoulders to get him through.

Eventually the poor sod ended up in BMH, his injuries, slightly bruised thighs and two broken wrists that had jammed on the lower hatch coaming.

Towny

In conventional diesel electric boats one of the most difficult and potentially hazardous evolutions was snorting, or snorkelling, travelling at periscope depth driven by diesel engine power, combustion air being drawn down the raised, induction, snort mast and exhausted through individual group exhaust valves.

Emergency shutdown from snorting is critical as everything happens at once. One of the most critical actions of this evolution is to shut the group exhaust valves to prevent water flooding back into the engines.

To this end, at snorting stations, two junior Engine Room personnel are stationed opposite each of the group exhaust valves that are fitted with large chrome plated hand wheels. When the stop snorting alarm goes off these two individuals have to leap across the gangway and spin the hand wheels to shut the valves pronto.

I had the misfortune to serve as an Engine Room watch keeper with a stoker who was a towny, (came from same town) and whose snorting watch keeping station was the starb'd group exhaust valve

He had one unfortunate bad habit in that he constantly picked his nose, and could produce some fearsome grollies. He also had a malevolent streak and got bored very easily, so to relieve his boredom and amuse himself at the same time he parked his grollies on his oppo's hand wheel in the hope that the stop snorting order would come through when the hand wheel was fully loaded.

Air Leak

Ever since I'd joined this 'P' Boat there had been a compressed air leak behind the main panel in the control room where the planesmen were stationed. The only time it became apparent was when we were at silent routine, when carpets were laid on deck plates, fluorescent lights switched off because of a slight frequency buzz, and all crew off watch turned in, and then you could hear this quiet whistle/hiss. It was considered not bad enough to warrant repair because it would mean a major strip down and a right bunch of buggers behind the panel until a bad electrical fault behind the same panel forced the issue. So the next time we were alongside *Adamant* the panel had to be stripped out, so it was decided to repair the air leak at the same time as the Greenies did their own thing.

A Coppersmith was sent down from inboard, I hadn't met him before, but I showed him the job and he seemed quite confident and competent and went back inboard to organise his gear while I isolated the offending pipeline and drained down the high pressure air from the defective section. The pipework was cupro–nickel–iron thick-wall 1½-inch bore connected together in this case by a capillary brazed union, which is a four-inch length of thick wall tube of the same material which is bored to a good fit over the main pipework. This union has recesses turned inside at both ends and filled with fluxed brazed metal (a similar system to fluxed solder in Yorkshire fittings, but with a much higher melting point). The two ends to be joined are parted and a sleeve slid on one end, the two ends are then lined up and the sleeve slid back halfway so that it is centred over both ends. Heat is then applied, the braze metal melts and fuses the sleeve to the parent pipe work and the joint is complete and tested when cooled. Unfortunately our leak was a flaw in the fusing of the union in one little area. The solution was to cut the existing sleeve in half, part the two ends and heat them up one at a time and slide them off. Then clean up the ends, slot on a new capillary brazed union, realign the pipe ends and as already explained, slide the sleeve half and half again, apply heat, cool down and test.

Our new Coppersmith seemed to still know what he was doing and had organised asbestos millboard and cloth (this was before the asbestos who-ha) to protect the surrounding area from heat. I'd cut through the old leaky union, he applied the heat and slid off the two halves no problem, cleaned up the ends, slid the new sleeve on, applied heat, let it cool down and I went to test it. High pressure compressed air is one of the mainstays of a Boat, and is stored in external (outside the pressure hull) bottle banks and piped into the Boat via high pressure hull stop valves. Compressed air at 4,000 psi is both powerful and dangerous if mishandled and very unforgiving.

The hull valve to allow air back into the repaired section was well forrard by the forends hatch. I left the Coppersmith in the control room and went up front to open the hull valve. You have to be careful when refilling an empty line, you have to open up slowly to allow the pressure to equalise slowly. A loud hiss tells you that you are bleeding air into the de-pressurised section and as it diminishes you open the valve a little more until eventually the hiss diminishes to silence indicating that the pressure has equalised, so I opened the valve fully and then heard that all familiar sound of an air leak but quite loud this time, so I started to shut the hull valve. As I just about had it shut there was a loud bang and a thick cloud of asbestos dust flew up the passage enveloping everything. All the crew converges on the Control room and as the dust settled there was no trace of asbestos left at the repair site and no sign of the Coppersmith either, and we all feared the worst. Eventually he was located just coming round, but very dazed with his upper body bare and pebble-dashed, lying on the wardroom door which was lying on the wardroom table, with his oxyacetylene torch still in his hand. What had happened was that as I'd opened the Hull Valve he'd heard the leak and without thinking had re-applied his torch to reheat the joint as the pressure had equalised at 4,000 psi the braze metal had melted and the built up pressure had exploded out.

Inspection of the pipework showed that the two ends had been blown apart a foot due to the reaction of the expelled air and no amount of heaving with chain blocks and pull lifts could get the two ends back together and a foot length of pipe had to be fitted with two capillary brazed unions. The lessons to be learnt, don't mess with HP air, it may damage your health or Coppersmith, also your ego and credibility.

HMS NUC. TRAINING *SULTAN*

Concentration

An incident on my 'O' boat reinforced my belief that lack of concentration could cost lives but could cause the loss of a whole Boat.

When a Boat comes in from an extensive patrol there are usually plenty of defects both minor and major and it was usual to send down assistance from inboard (Depot Ship) to help out, and for all fitting work spare crew personnel are used as they are qualified Boats Tiffs.

One particular individual, who shall remain nameless, obviously had something other than work on his mind. The job he was given was to replace a slightly leaking hydraulic actuator seal on number seven Diving Main Vent, the single one right back aft. The routine in any job on Main Vents is to ensure that the boat safety is paramount and to ensure that the vent is fully shut, cottered and locked off before commencing the job. The cotter is inserted to ensure that the vent is mechanically locked shut and with it fitted not even the hydraulics can open it. Next the hydraulic supply and return of the actuator have to be isolated and any residual pressure which is locked in vented off, which as will be remembered from science at school is easy, because liquids are virtually incompressible, so one little drop released and the pressure all gone. The Fitter decided that as the seal was leaking he didn't deem it necessary to do it, as he assumed that it had already depressurised. If he had followed procedure with this type of repair, none of the following would have happened.

The one major thing that our fitter failed to do was to vent off the air side of the Air Loaded Accumulator. These units are fitted to any hydraulically operated unit that maintains hull integrity and are fitted so that they cannot be isolated from the attached unit. It is fitted to ensure that in the unlikely event of a loss of hydraulic pressure the Air Loaded Accumulator, at 4,000 psi, will take over and ensure that the item it is connected to stays in its appropriate safe position. In the case of a main vent, shut, as it has, as its name implies, air pressure on top of a quantity of oil. The correct procedure at this stage would have been to drain the Air Loaded Accumulator which unfortunately was omitted.

To progress the repair the Fitter had to disconnect a hydraulic union, which on this Boat was sealed with polymer 'O' rings. If the main 'O' ring starts to leak a small hole in the union nut starts to weep hydraulic oil and indicates that it has to be repaired at the next available opportunity. Our Tiff gaily started to undo the union nut, but 'O' seals are notorious in that they give way suddenly when it is undone with pressure behind it. The pressure, as already mentioned, is 4,000 psi (four thousand pounds to every square inch) which is the equivalent of 20 large men all standing on an area less than the size of a fifty pence piece.

As we inevitable the 'O' seal eventually gave way and the hydraulic oil shot out of the tell tale hole in a vaporised stream. It could have hit the bulkhead and made a mess, it could have hit the Tiff and given him an embolism with possible fatal consequences, but sods law dictated that it went straight into the centre of a switched on electric fire and immediately burst into a roaring inferno, which was fortunately extinguished in record time by alert crew members, but caused extensive damage.

The perpetrator after a full enquiry was returned to General Service.

Tot Time

When nuclear power first came on the RN scene I was determined to be in the programme as soon as possible, and slapped in immediately to attend the next Nuclear Course. My requested was eventually granted and at the appropriate time my draft chit came through and I was off down south to join HMS *Sultan*, the RN Engineering School.

A dozen of us started the course, all of us technicians of one sort or another, ERAs, EAs, Mechanics and Electrical Mechanics. It was all very different to what we'd been used to and it was all classroom work, and very hard too. It was long before calculators were available and we had to be taught to use slide rules before we could start on any other subjects. Subjects like higher maths, calculus, water chemistry, reactor physics and Pressurised Water Reactor design, to name but a few. We even got homework to do, and many an unhappy hour was spent wrestling with a formula as long as your arm to resolve a reactor design problem. It was very hard work with no let-up apart from one amusing incident that involved me.

I was one of the few on the course who lived aboard *Sultan* so was subject to weekend duties, a pain in the bum but unavoidable. Having three buttons (Chief) one of my duties were witnessing the troops' rum ration issue. This particular Saturday the rum ration came up spot on time at twelve noon, with all the troops left aboard queuing up with their tongues hanging out. It was my duty to witness the consumption of their tot on production of their station card. The tot was an eighth of a pint of hundred proof dark Demerara Rum mixed with a quarter of a pint of water, known as two and one, that had to be consumed on the spot, which was the reason for me being there. All went well for a while, but after a short while I thought I recognised one of the sailors in the queue. He showed a station card, got his tot, turned away from me and downed it in one and sort of slid off. It got me thinking how many more were at it, and dawn me if meladdo wasn't back again ten minutes later for yet another tot, which was taking the urine. I let him get near the issuing table and then called him to one side and told him, in a very quiet voice that could not be overheard, that two tots was one too many and if he didn't piss off and take his cheating mates with him I'd troop the lot of them. The results were dramatic, as he walked slowly down the queue almost half of them peeled off and disappeared.

Divisions

The Nuclear course continued at breakneck speed and we were all hard pushed to keep up, but one advantage of us not being RA was that one of the instructors, who was also RA, had completed the very intensive United States nuclear course, but had not been able to join the first nuclear boat (*Dreadnought*) due to ill health. John Redwood was a Chief Tiff and also an outstanding instructor who lived aboard like us. Being Tiffs we shared the same Mess and John was always on hand to give us help, but he had an awful problem, or should I say we had a problem with him!

He looked a lot younger than his years, in fact he was near the end of his 'fifth year'. He had served with distinction in the Second World War, but not in the mob, in the RAF and he had the medals to prove it. For this reason nobody wanted to know him when we were on Sunday Divisions because if you were to stand anywhere near him the Inspecting Officer would stop at John and suggest that he had the wrong medals up, so John had to explain that he had been a rear gunner in Lancaster bombers in the last war, which was all very well for him as they had a good natter and he was congratulated. The snag was that the Inspecting Officer's entourage ranging down the line were staring at you trying to pick holes in your turn-out. At the order to fall in, blokes would pull all sorts of strokes to be as far away from John as possible.

Blackout

Before I start this tale of woe I must explain that *Dreadnought* was a one off, the reason being that the forrard end was British design, but aft of the Reactor bulkhead was pure American, the Skipjack class.

So forrard was all crew accommodation, torpedo stowage and torpedo discharge tubes, galley, domestic tankage (fresh water, slops, sewage, batteries), Wardroom and of course the Control Room, while back aft was the all important Reactor compartment with its associated Primary Circulating System and the steam generator. Aft of which was the Auxiliary Machinery Space (AMS) with all the Reactor System electronic controls. Further aft were the Manoeuvring Room, fresh water distillation unit, main AC generators, twin turbine main engines and coupling gearbox to the main propeller drive shaft.

Situated in the AMS lower level lurked a bit of a beast. It was a large diesel engine, the type used as main engines on the old American Guppy Class conventional Boats. It was shackled to a generator, and was there to be used as an emergency power supply should all else fail, sort of last ditch get your home system!

As an AMS watch keeper at the time, one of my duties was to take my turn at flashing up the said Diesel Generator which was not too difficult, but had to be prepared for starting very precisely.

The engine was quite unique as to get maximum power whilst taking up minimum space its design was unusual. It was a vertical, twin crankshaft, opposed piston, two-stroke, and very compact. The start routine had to be strictly adhered to, duel system primed, cooling water system filters checked clear and opened up, compressed air start system fully pressurised and finally the forced lubrication system priming pump run for exactly four minutes immediately prior to start up. Any less time and some upper crank shaft bearings would be dry, any more and excess oil would collect in the air intake duct and cause heavy smoking on firing up!

I had just relieved my oppo and taken over the morning watch in the AMS. He passed on to me that the Manoeuvring Room wanted for exercise, the Emergency Generator Engine to be run up. He explained that he had checked all systems for me, and went off to get his head down as he was nif-naffed!

I, to be certain, rechecked that everything was OK, checked with the Manoeuvring Room that we were ready to do a run it up and got permission to carry on and told my Stoker to start the lub oil priming pump. Timed four minutes exactly, stopped the priming pump, pulled the start lever, the engine started to turn over and gain revs. The next thing I remember was an almighty bang and I'm blown back spread-eagled on the auxiliary pipe work and can't see a thing for acrid black smoke.

My first reaction was to check that I was still in one piece, the next to shout out to check that Stokes was OK and got a choking reply, the next to wonder how much my slop chit would be?

When the smoke finally cleared and we checked round we found that all the handhold doors in the induction manifold had been blown out and stokes and I had been very lucky that they had missed us as they had blown out with such force they had been almost completely destroyed.

The inevitable investigation and enquiries were carried out and I'm thinking about demotion and charges, but when all came to all it appeared that the bloke I'd relieved failed to pass on to me that he'd already primed the lub oil system so it had had double whacks. The excess oil had collected in the intake manifold and the first pot to fire had flashed back and ignited the excess oil with dire results. I was exonerated and my oppo was given a reprimand and we both had to straighten and repair the hand hold doors and remake the beyond repair securing strong backs, in our own time of course!

AMS

HMS *Dreadnought* was doing a dived transit and all was well and quiet. It was an ideal situation with all watch keeping positions being fully and well covered apart from one. That one was the Auxiliary Machinery Space lower level, or the Stream Generator Control Panel.

The system works like this, the Nuclear Reactor generates heat which is absorbed by very pure water which is then circulated round the Reactor Core by very large pumps. Very high temperatures can be achieved because this primary system is run at very high pressure to prevent departure from nucleate boiling (DNB) which would be catastrophic. This primary water system at high temperature is circulated through a very large heat exchanger, known as the Steam Generator because as the name implies it produces steam to feed the Engine Room main engines and generator turbines. This occurs because the secondary system runs at a much lower pressure than the primary system. The secondary system therefore has to be managed very much like a normal boiler in that the water level has to be controlled and maintained, and the water quality also has to be monitored and maintained to very close tolerances which is where the watch keeper on the control panel comes into his own. Water feed pumps have to be monitored, water samples taken and analysed and the appropriate chemical treatment administered to maintain water and steam quality.

This watch keeping position is therefore a most important and responsible spot to fill. The snag was that all was not well as one of the said watch keepers was less than keen and responsible. It wasn't that he couldn't do his job, it's just that he has a terrible time keeper.

One of the cardinal sins in the RN, especially so in boats, was being adrift so being adrift relieving your watch keeping oppo was even more heinous.

This particular individual, a Stoker PO who shall remain nameless should have known better, but didn't seem to give a toss as he was always adrift and not by small amounts either.

One ritual with the Tot was that it usually came up well before watch change at noon, so one of the forenoon watch keeper's oppos would relieve him for a couple or so minuets to get his tot and get back fast. But not me-laddo, if he was relieved at tot time he used to take his time and sometimes even forgot that he was still on watch, it just illustrates how uncaring and unthinking he was, but this was about to change for the better.

This particular incident was not a tot time event but a standard watch change over where the relieving bloke would usually turn up at least ten minutes early, or even more, so that a good hand over could be made and any relevant information passed on. One of the Steam Generator watch keepers was taken ill so my mate a fully qualified Chief Tiff who was day work took his place on a temporary basis. This particular time he had the Afternoon Watch and while he was there and things were running smoothly he had taken his tool bag down with him and did a small repair job on a leaking sample line and guess who had the First Dog to relieve him, yes, me-laddo who turned up a full quarter of an hour adrift!

My mate was seething inside but kept his cool, just. His relief eventually turned up with a tin of Coke in his first, and with no apology or kiss me arse sat on the bench alongside my mate passed him the tin of Coke, said that the queue at the canteen had been too long and the service crap and he'd forgotten his can spanner and would my mate open it for him?

My mate, now absolutely bubbling inside said with a very strained smile, 'certainly', fished in his tool bag, took out his biggest screwdriver, laid the can on its side next door to its owner and with one swift stroke stabbed it straight through, lifted the impaled tin and offered it to the owner who, totally gob smacked withdrew it from the blade and got absolutely blathered in very lively Coca-Cola. My mate then read his horoscope and I'm sure he got the message because he drastically changed his ways!!

Butlins

We ended up in Singapore and were billeted in HMS *Terror*, the Butlins of the Far East.

We were as green as grass when it came to living ashore in that climate so some of the barrack stations took us under their wing and showed us the ropes.

The weather was hot but very humid, so we couldn't get our heads round the fact that all the wardrobes had illuminated light bulbs in the bases day and night. One of the old lags showed us a suit that had been in an un-illuminated wardrobe and it was green almost all over where food had been spilt. In the humidity mould grows at an alarming rate overnight.

Drinking in the mess could be a nightmare if you weren't careful because you were always thirsty with the heat, so you went down to the bar and got a beer, Tiger beer, bottles of it, and sank the first one in one go and suddenly you had excruciating stomach cramp because the beer was straight out of the fridge and damn near frozen. You learnt after the first time to take it slowly to start with.

Eating out was another experience not to be missed. One of the less affluent streets was all just shops and cafes combined. One the pavement outside the shop was the Chef and his kitchen with his wok and charcoal brazier with all his makings around him. You go into the shop, which sells all things edible, and go upstairs to the cafe area and we always used to go up to the front so that we could look out. You were brought a large pot of green tea and cups at no charge and asked if you wished to eat. If yes the menu was brought you made your selections and ordered and then you leant out of the window, which was not glazed, and watched the chef preparing your meals. His utensils consisted of a razor sharp cleaver and a soup ladle with which he performed miracles and the food was superb.

My favourite dish was called by 'Jack' stir fry Steamroller Duck. It was preserved duck in the shape of a duck complete with skin but no bones in it, and had been preserved with the long neck formed into a hook shape and pressed between two flat boards. When the process was complete it came out the shape of a duck, but only half an inch thick, which is why Jack called it Steamroller Duck, for obvious reasons. They were displayed by the Chef hung on a rail by their necks hence the shaping.

If you wanted to eat in the evening a good and very interesting place to go was Bougiss Street. During the day it was a normal thoroughfare but in the evening the tables and chairs came out onto the street from the cafes and restaurants. My favourite dish there was spicy king prawns where half a dozen filled a big plate on their own and for me was a full meal with nothing else.

The place was a hotbed of theft, corruption and prostitution where three and four year old little girls would play you at noughts and crosses for money, you could if you were lucky draw, but you could never win.

Then there were the Ky-Tyes, Lady Boys, we'd been warned, but some hadn't. These creatures dressed in the latest revealing fashion dresses, and acted and looked far better than the local girls. You could see the young Merchant Seamen making a play for them, and were going to get a nasty surprise, like a handful of nuts and bolts.

Danny, a mate of ours had a potentially difficult problem. He excused himself and trotted off to avail himself of the nearest urinal. Halfway through he was jostled, so checked his back pocket and discovered that his pay book had been lifted and out of the corner of his eye saw this native scooting for the exit. Danny was built like a brick toilet and very fit so took off in hot pursuit. The thief was very quick dodging round the tables, but Danny had a better idea, he ran over the tables with plates and bowls scattering all over the place. He caught up with the thief and launched himself on top completely flattening him and recovered his pay book which was just as well as loss of said item was a heavily punishable offence!!

A Good Run Ashore

We'd been day running out of Gibraltar which meant that one watch per day could have a full day inboard and have a good run ashore.

The snag was that me and a couple of others of the wreaking gang had to stay aboard full time because the outside hydraulic system had been contaminated with sea water via a Periscope Ram seal leak. A small centrifuge had been shipped out to us from UK and we had to rig it up in the forrard AMS to distilled water wash the hydraulic oil and centrifuge the resultant emulsion to remove the salt from the system to prevent corrosion to the hydraulic system internals. After a gruelling fortnight of this day and night we managed to crack it and the Engineer swung it for us to have a whole week's leave.

We found a small Hotel just off Main Street, cheap and clean, and near to our favourite Boozer come eatery 'La Campana' (The Bell), again just off Main Street up Engineers Lane. We knew it as Eddies Bar, and we could be found each morning sitting on his front door step waiting for him to open up. When he did, his first move was to put a bottle of Bacardi and a few cokes on the bar as Freebies, after that was consumed we paid. Up a couple of steps to the rear of the Joint you could get a good meal. Eddie was real good to us, he even made us special ice cubes that would fit down the neck of an ordinary Vacuum Flask which we then topped up with Bacardi and coke and a few twists of lemon. We'd have a few more bevies then wander down to the beach, buying a load of fresh fruit on the way, do a bit of sunbathing, eye up the ....??????????

This particular day though was different, it was one of the lads birthday and we went a bit OTT, culminating in the birthday boy suffering a dislocated shoulder. He'd accidentally fallen between two parked cars and was laid on his back between them singing quietly to himself. His oppo went back to assist and grabbed both wrists to pull him up. The snag was that the cars were parked quite close together, and his oppo's shoulders were wider than the gap with the inevitable result, but he didn't find out until the next morning he was that handcarted. We all gravitated back to Eddie's Bar to get big eats, and had a lovely seafood salad. The birthday boy, managed to eat with one hand, complained that the battered onion rings were rubbery. His oppo called his a stupid b—.

'They ain't onion rings, them's octopus testicles.' He was famed for his malapropisms, and was widely quoted as having the following diatribe extolling the virtues of his local GP.

'By eck, e is a clever fella, real clever, e's got all these Diplomats angin on its wall. An e's got all this Tubeless furniture in is waitin room. An e's got a couple of bloody big butiful Laburnam dogs in is back yard. E's reel clever, e's even sown a new lawn an' its cumin' on a treat 'cos e's used Bird Propellant on it!!!'

Smart Arse

Electrical Artificer Buster Merryweather, built like a brick toilet with a face to match, but a very clever and competent Tiff who was top notch at his job. His main function being to look after our state of the art Asdic gear.

One of his minor duties was to service and maintain the depth recorders, of which there were a few. One was an upward pointing unit to record our actual depth, and also the thickness of ice when in those climes, so being a very important unit.

During a transit this unit became US and was reported to Buster who came up to the control room to fix it. I was on watch on the panel and heard a lot of swearing and cursing coming from the other side of the Control room, so as it was quiet, I wandered across to see what all the commotion was about. There was Buster trying unsuccessfully to prise an electronic bottle, about two inches in diameter, out of the face of the unit. The end of it was flush with the face of the recorder and about six inches long. He'd been trying to prise it out with a pair of screwdrivers by digging holes in the sides of it and levering, but the thin aluminium casing just kept splitting through the front and he was getting desperate and very frustrated. I volunteered to get it out for him, he looked a bit unconvinced but agreed to let me have a go, so I called over the watch Stoker and asked him to go forrard and get me a couple of yards of cod line.

Buster asked me, rather scornfully, what I intended doing with a big of string, and I told him to wait and see. The stoker returned with the cod line and I formed a timber hitch in the middle of the line by forming two loops and putting the second one in front of the first, and slotting the two loops over the butchered end of the bottle. I borrowed Buster's screwdriver and pushed the loops about two inches down the tube, pulled both ends tight with a bit of jiggling then took a firm hold of both ends and have a hefty yank and out popped the whole thing. Buster's expression was a mixture of disbelief, surprise and relief, he turned to me and said:

'Cor, thanks Ted that's great, but nobody likes a smart arse!!!!'

Getting Your Own Back

Their Lordships in their wisdom decided that as we, the first Nuclear Boat, was at sea for prolonged periods our dental care was suffering and so decided to lash us up to a Dentist to augment the Medical and Health Physics department.

Therefore on our next call at Faslane a very young Sub-Lieutenant joined us, together with a whole host of tools and equipment. Me and my oppo as Outside Wreckers were detailed off to rig his gear and install him in the Health Physics Lab. This involved the securing of a rather large dentist's chair and supplying it with power and plumbing, and more interestingly a state of the art (at that time) Air Turbine Drill which as its name implies required clean compressed air and a cooling water supply.

The Air Turbine was a wonderful gadget which made tooth drilling much less painful and uncomfortable. The only snag was that it had three minor drawbacks, to start with it made a really high pitched whine, secondly it produced an acrid smell like burning plastic, and finally the water to cool the cutter and the tooth had to be sucked away with a vacuum pipe which was hooked into the patient's mouth. All this at that time was quite revolutionary.

Sub Lieutenant Andy was not only good at his job, he was also a good run ashore with the lads, which tended to be frowned on by the Wardroom, but he didn't give a toss as he was only on a Short Service Commission.

Eventually the crew's teeth were deemed to be one hundred percent and Andy was to be drafted elsewhere, so a few of us Senior Rates took him on a final run ashore, got him well smashed, smuggled him back aboard and down to the Health Physics Lab and, to his surprise, tied him in his own chair.

My mate and I having installed all his gear also knew how it worked and how to use it. Poor sucker Andy thought it was all quite amusing until we prised his mouth open and kept it wide open with an expandable clamp, then he started to change colour a bit. My oppo got one of his tooth band clamps, the sort used to contain filling amalgam material until it hardens. This time though, unbeknown to our victim, we used it to clamp a piece of Perspex which we had prepared earlier, to one of his incisors, then I got to work with the air turbine At this stage the poor sods colour completely vanished and his complexion resembled alabaster. Then I started grinding away at the Perspex which was in close contact with his tooth which made it appear that his tushy peg was being whittled away. He was now acting like a sweaty chameleon as his colour had changed to a pale shade of yellow. I redoubled my efforts on the Perspex, causing the burning smell to become more pungent and steam wafted past his now saucer wide eyes with an accompanying change of colour to puce. We finally called a halt and unshackled the Perspex and took the jaw clamp out, but by this time he'd had a complete humour failure, but the relief on his face was a picture. He did eventually forgive us, but he admitted that at the time he had been scared shitless, and clewed up stone cold sober.

Loose Talk Costs

My mate Knocker was Duty Tiff on *Dreadnought*, and I just happened to be passing when the Shore Telephone rang in the Control Room. Knocker answered it and after a short conversation called me over and asked me to do him a sub as it was urgent. I hadn't anything planned so I agreed, telling him that it would cost him, to which he agreed and then shot off towards the Mess. The next thing I remember was Knocker, in no time at all, leaping up the forrard hatch and away ashore, all dolled up and smelling gorgeous.

When a Boat comes in from a long patrol or exercise there are usually a few major defects that require inboard, Depot Ship, assistance. This particular time some Brazing was required and 'Jumper' Collins came aboard with his gear to do the job. He was a good hand, for a General Service Tiff, and one of the best coppersmiths around. All the Boat's Senior Rates knew him from previous jobs, and he was always invited into the mess for a Tot and Lunch.

Knocker his tongue loosened by his tot was telling, no bragging, about the previous evening's evolution.

'I was on duty when the shore phone rang in the control room. I answered it and a lovely Female voice engaged me in pleasant conversation, eventually getting round to telling me how lonely she was and that

I sounded nice, and that if I fancied I could call round with a bottle of Whiskey and see what developed. So Ted did me a sub, and she gave me her address, a block of flats in Dumbarton, I got cleaned up, threw on some civvy lagging, got a bottle of Famous Grouse and went up homers. We saw the skey off and had a hell of a night. I've been invited back again tonight.'

Lots of catcalls followed with crude and other type comments. The subject eventually changed and a few other topics were discussed during which Jumper was heard to ask Knocker which block of flats it was in Dumbarton because he thought he knew who it was.

'It's McAllister House and there's no way you could possible know her.'

'Don't they call her Jean Logan the girlfriend of the Scratcher of that "O" Boat that's just left on patrol?'

'No, they call her Craig and was unattached until last night.'

'Oh sorry, my mistake.'

And so the conversations meandered on until turn to time.

That evening Knocker got tarted up again and literally floated ashore, he was so hyped up. Two hours later he was back aboard, fuming and threatening GBH. Eventually he'd calmed down sufficiently to tell us what he was so wound up about.

'I got me bottle, caught the train to Dumbarton, went up to the flat, rang the bell, the door opened and there's that snide bastard Coppersmith with a half empty bottle of whiskey in his fist, and tells me to f— off and slams the door in my face.'

With that the whole of the mess collapsed in a big heap of mirth.

'Jumper always was a quick flip to the front, how do you think he got his nickname?' someone chortled.

'That'll teach you to keep your trap shut Knocker,' another threw in.

For those who are a little confused, in Scotland, flats in big cities are very common and have an intercom system with the occupants' names against flat numbers.

I'll Take It Through Your Ears

A good mate of mine on 'Dreaded' was Bunny Warren an REA (Radar Electrical Artificer) who joined boats to specialise in sonar, who looked more like a prop forward than an electronic whiz kid. The set he looked after was, at the time, so good and so secret that the Yanks would have given their eye teeth for it. It would do a series of pings, process the return echoes, photograph the results and back project them onto a one metre square screen every 45 seconds where a couple of sailors would mark in chinagraph pencil possible contacts and erase them if they did not develop. Their Lordships in their wisdom also required a 35 millimetre record of the same screen, so I was given the task of mounting the camera in a suitable position, which I did on the Bulkhead to the rear of and between the two watchkeepers. I also rigged a remote shutter release, so that all the operator had to do was his normal job and then check the camera's view was clear and press the release as it had automatic wind on. Unfortunately a newly qualified Sub Lieutenant had joined us as a spare hand and took a particular interest in the sonar compartment and would spend all his spare time peering at the screen. One particular operator, a laconic TGM from the Outer Hebrides always seemed to be on watch when the nosy Subby was in attendance and was always in the way when the photo had to be taken. Eventually he got so fed up of asking the Subby to please excuse him and have him apologise that eventually as he turned, with the Subby in the way yet again, he held up his hand and said in a loud voice:

'Don't bother moving Sir, just stand sideways and I'll take it through your f—ing ears.' Exit very red faced Subby to ribald laughter, never to be seen in the sonar shack again.

Gear Wheel

'Dreaded' being a proper Submarine rather than a Submersible we were in great demand for exercise purposes with all the NATO fleets, especially with our sonar set to help us.

An incident I was involved in which gave me great satisfaction was when, just after one such exercise had started, our super duper sonar set packed up. I went to see Bunny Warren our REA who told me that an aluminium gear wheel driving the micro switch sequencing shaft had stripped some of its teeth and was inoperable so the Skipper was going to have to call the whole exercise off, so I talked Bunny into letting me have a look. Having seen the damage I reckoned I would stand a fair chance of effecting a repair, so we had a word with the EO who in turn had a word with the Skipper who reckoned that we had nothing to lose and to go ahead. So it was back aft to the upmarket lathe where I turned up a gear blank the same size as the existing damaged one in brass. I cut out the damaged section of the gear wheel and cut and fitted a brass section of the blank to fit the gap then pinned and epoxy resin glued it in. Then came the tricky bit, re-cutting the teeth as it was helical gear, but carefully use of needle files saw the job done, Bunny and I refitted it back in the unit and gave it a quick whirl and lo and behold it worked and the exercise went ahead after a short delay. The sequel to the story is that my recommendation that future units be fitted with gearwheels manufactured in yellow metal instead of aluminium was adopted and I won an award of a glass bottomed pewter tankard through the Herbert Lott Trust Fund (or as we called it the Herbert Trot Lust Fund). When it finally turned up along with a few other blokes' awards and promotions we were all called into the wardroom where the Skipper did the presentations. When it came to my turn he opened a can of beer and said he wanted to christen it for me and poured it into my brand new tankard and straight onto my boots and the wardroom carpet so I had to go away and mend the damned thing, the story of my life!!

Shut Up

One amusing incident happened during the first dog watch in the control room. We had been having trouble with the fresh water tank automatic change-over system which was not working because a sensor had malfunctioned and we didn't carry a spare so we wouldn't be able to fix it till we returned to base. The one good thing about a Nuclear Boat was that fresh water was really quite plentiful and was stored in two relatively large tanks, one being in service while the other would be either full waiting to go into service or being filled by the back aft distillation unit.

With the automatic change-over system out of action it was PO Stoker Steve Cutt's job as tanky to do it manually.

All tank operating movements are performed using compressed air so that when Steve changed over fresh water tanks his routine was to depressurise the on line tank, divert the distiller output to the now depressurised tank and isolate it from the supply system. The tank has now to be isolated from the distillation line, opened up to the service line and then pressurised to give adequate supply pressure at the taps. It sounds complicated but in fact it takes no longer than it takes to read this explanation.

Our Skipper at this time was one of the best gaffers I've ever served under and shall remain nameless as he is no longer with us. At the time of the incident he was a very senior three ringer, not too tall and rather portly to say the least. I was on watch on the diving panel on a deep dive transit, so everything was quiet and everyone was alert but relaxed when the fresh water low level alarm went off so I called Steve on the intercom to have him change over the tanks. Steve, who was also not very tall and also portly, came up into the control room to do the biz on the tank system.

He had just depressurised the nearly empty tank when this apparition appeared at the entrance of the passage leading to the wardroom and officers' bathroom. All eyes in the control room focused on the figure of the Skipper with a flowery bath hat on, a skimpy towel round his ample waist, covered from head to foot in very heavy layer of soap lather. Poor old Steve couldn't contain himself and burst into a very loud uncontrollable laugh. Our beloved Captain stood in all his glory, now with a bright red, almost purple face and exploded with the words:

'Shut up, fat bastard!!'

And the whole of the control room collapsed in a big heap of hilarity whereupon the Skipper spun on his heel and, as dignified as possible, marched off back to the bathroom with the noise of very loud laughter ringing in his ears.

Poor Sod

I met an old mate of mine from apprentice days. He had made the Piggery (wardroom) but tended to be still one of the lads. We went for a run ashore in Helensburgh, had a few bevies and a good natter. He came up with the story of a guy he had served with.

This chap had recently qualified, and at the time of the incident was Officer Of The Watch in the control room on a dived listening transit which was very boring. To liven things up the rest of the wardroom (negative Skipper) decided to wind him up a bit, but unfortunately it went a bit further than expected. They knew that he was fluent in Russian and Morse Code, so one of the perpetrators who also had Russian wrote out a bogus message listing Soviet Naval Manoeuvres and passed it on to the Communications Officer.

They then went back aft to the shaft space which is where the propeller shaft disappears into the stern gland. Also down there is the plumber block and thrust bearing which transmits the engine power to the propeller. The shaft is also fitted with a wrong direction transmitter which as the name implies is that if the manoeuvring panel operator answers the telegraph in the wrong direction a big red light comes up on the steering and diving panel. The swine's down the shaft space started sending their pre-prepared message in Morse on the wrong direction transmitter.

Said unfortunate mug took it in hook line and sinker, started reading the flashing red light, grabbed a pencil and pad and started scribbling, and on completion he grabbed the intercom mike and in an urgent voice called, 'Captain to the Control Room.' Out came *mon capitaine* looking a bit peeved at having been disturbed, and went into conference with the victim, his face started to distort and change colour, suddenly he grabbed the intercom mike and yelled into it for an immediate relief for the OOW.

My mate shot out of the wardroom, thinking it was an emergency and took over the watch, but much to his dismay and amazement dragged the poor gullible unfortunate into his cabin, whence the verbal volume was horrendous.

After the Skipper's minute investigation the perpetuators were severely reprimanded and lost seniority, the victim was given a reprimand and a draft which is where I came across him. I was wrecking on 'Dreaded' and had just been redeployed to back aft to train up for watch keeping duties. He had just qualified as OOW in the Manoeuvring Room where he lacked confidence and was of a very nervous disposition. The Engine Room watch keepers were awful to him as Jack was always keen to exploit any weakness. The rotten sods made his life a misery, much of it, self inflicted. For instance, on start up when steam is first exported back aft, all steam lines become waterlogged (full of water) and the automatic steam traps could not cope with the sheer volume, so trap bypass valves were fitted which were manually operated with great care, apart from the one in the far corner just outside the front of the manoeuvring room, which the bloke operating it wangled wide open after giving the appropriate warning signal to the rest of the manoeuvring room gang. The result was a horrendous roaring of escaping steam, with spectacular effect on said OOW who would leap about a foot in the air trying not to panic.

Another rotten trick that was inflicted on him was the Kensitas scam. Kensitas cigarettes used a bright red paper strip to pull out the first fag from the flip top packet, it's not surprising how boredom can bring out the worst in Jack. The idea was to remove the red strip, lick it, stick at an angle on the cheek, stick the fingers, well splayed so that the red showed through well, then with an ear splitting scream stagger past the manoeuvring room door groaning loudly. This brought me-laddo almost diving out of the room in sheer panic to comfort a junior stoker who was grinning like a Cheshire cat. By this time the object of all this effort had developed a rather nasty nervous twitch, and an even nastier looking skin complaint which required medical treatment to his hands and plastic gloves to protect them.

The last straw was another cruel trick that the Engine Room watch-keepers pulled off on him which involved the emergency manoeuvring position. On Captains rounds he always praised the Engine Room crew for the areas cleanliness, apart from the area behind this position which was inaccessible due to two big red handwheels which were meant to be used in the event of the manoeuvring room throttle control position being disabled. They were situated tightly up between the main turbine casings, so the stoker whose part of ship it was, asked the Tiff on his watch to undo the nuts securing the hand wheels to their respective spindles, high irregular but needs must. The Tiff duly obliged, and the stoker was able to remove both hand wheels and squeeze down the resultant gap to clean the offending area. This routine was passed on to the stokers of the other watches and the said area was kept pristine, much to the Skipper's and Chief Engineer's delight. The only problem was that the securing nuts were left hand tight, which was not really good practice.

Nevertheless, some bright spark thought up another spiffing wheeze to terrorise said nervous wreck. Again in was in the silent hours, dived in transit, all very steady, quiet and boring. He unshackled the hand wheels and took the ahead one (the biggest) and, on command the whole Engine Room watch started a hell of a racked by banging hammers and wheel keys, the perpetrator booled the hand wheel past the manoeuvring room doorway. The target literally fell out into the passage in sheer panic to view the grins of the whole of the Engine Room watch.

He left shortly afterwards, back to General Service (nearly a basket case). To Their Lordships it was obvious that he was not Boats material, and the kindest way career wise was to remove him medically, rather than label him generally unsuitable, which would have ruined his career in the Navy completely. In actual fact I'm told he went on in General Service to do rather well.

Run In

Me and my second dicky Chalky White decided to wind up our Chief REM, which turned out to be a bad mistake. We were laying alongside *Adamant* at the time kicking our heels and feeling very bored and all worked up.

The Chief Radio Electrical Mechanician in question sported a full set in the style of King Harken of Norway, a blond pointed beard and matching waxed moustache, which were his pride and joy. Normally when inboard he used to twirl his tash with soap which when dried was well pointed and quite impressive, but on runs ashore he waxed it into long rigid spikes. He'd been ashore and had a good bevy and had settled into an armchair in the chief's mess and fallen fast asleep. We arrived back aboard a bit later on after a good run ashore and espied said CREM and decided to wind him up a bit.

Chalky's hair was about the same shade as the target's so he got a bit of his forelock and twisted it with pussers hard (soap) then snipped it off. For all the world it looked like one side of the victim's tash. Chalky then did the dastardly deed, but at this juncture I must point out that said victim was known by all and sundry, except our dastardly double act, to be AC-DC, especially when in drink. Chalky, naively, with the whole of the mess members now watching him with great interest, crept up behind the mark holding the dummy tash in his right hand, also grabbed and tugged the real one to wake him up and then loudly snipped a pair of scissors in his ear and let go of the real one whilst displaying the fake. The mark leapt to his feet, saw the dummy tash and the scissors in Chalky's hand and suffered instant humour failure, in fact he went absolutely ballistic and chased poor old Chalky all over the place, threatening to do very naughty and rude things to him. Eventually Chalky found a hiding place in the Wardroom Pantry 'cos the Steward owed him a favour. The next day however, at tot time the CREM realised it was just a wind up and saw the funny side of it and had a good laugh with Chalky. The long term effect, though, was much different in that whenever Chiefy came off shore bevvied up and he knew that Chalky was about the hunt was on, threatening to inflict dire and dastardly acts of him, and Chalky, when he heard on the bush telegraph that he was being hunted, used to go to extraordinary lengths to hide as best he could, whilst I got away with it scot-free.

Ambition

My ambition in Boats was to serve on the first Nuclear Boat, *Dreadnought*, so I slapped in a Request Form to join the Nuclear Engineering course at HMS *Daedalus* down south.

To cut a long story short I passed the course and was drafted to *Dreadnought* as a part three, passed that and ended up as second dicky outside wrecker, trained up a replacement and ended up back aft, Engine Room watch keeping, so far so good!

Eventually I made it to my ultimate goal, the Aft Manoeuvring Room, starting off on the main engine control panel, which was a doddle as the only time anything really happened was entering and leaving harbour apart from the odd rev change.

The next stage was the electrical distribution panel, which was a lot more of a challenge. One of the trickiest operations being paralleling up the Turbo-Generators. It goes like this, the Engine Room Tiffs run up the off line TG and when happy with its performance, hand its control over to the gaffer (OOW) of the manoeuvring room who then tells the panel operator to do the biz. Then comes the tricky bit, getting the incoming genny to run at the correct speed. It's called the synchronous speed, which means that both gennies are running at exactly the same speed, which is achieved by the operator using a speeded switch which remotely adjusts the drive turbine governor and so the rotor speed.

The next tricky bit involves the synchronisation gauge which is like a clock face, but with only one hand which is adjusted by careful use of the speeder switch to rotate the hand slowly clockwise approximately once every five seconds. This means that the incoming unit is running minutely faster than the loaded unit, the breaker (big switch) used to bring the off line unit on in parallel is prepared by priming the breaker operating switch when the synchroniser gauge hand is at five minutes to twelve and rotating at the right speed. Now comes the REALLY tricky bit 'cos you have to do it all over again, the problem being that the minute speed differential is controlled electro-mechanically and is not always steady and the rotation if the indicator hand can become somewhat erratic. The snag is that this time the switch is operated the main breaker slams in and connects both TGs together, locking them electrically. If you get it wrong the best that can happen is that there is a slight thud and the lights dip momentarily and a cheer goes up throughout the boat, causing the operator some embarrassment. But if you get it badly wrong there is an almighty bang as two great lumps of machinery try to stop to phase lock together. It has been known for this type of incident to cause severe reduction gear box damage entailing extensive downtime and expensive repairs, and of course a very rough ride for the unfortunate panel operator.

I've seen trainees reduced to quivering wrecks trying to make the final switch, even having to be relieved by an experienced operator to show then that the trick was to take time and not worry if the delay was prolonged, as it was his responsibility and not to be rushed under any circumstances.

The final qualifying hurdle was the reactor control panel, which to the newcomer was awesome, the reason being its massive power potential, or to put it another way, destructive force if something should go wrong. For this reason the safety systems were extreme, such that all the control parameters were in triplicate and an automatic two out of three logic was operated by bank upon back of electronic racks containing row upon row of magnetic amplifiers, as this was the days before microchips. One of the snags with mag-amps was their unreliability due to temperature fluctuations, the solution being to install huge air conditioning units, which was, as a spin-off, also good for the crew, us being the last consideration.

Basically the system was that very pure water at very high pressure is pumped round the reactor core picking up a high temperature without generating steam, because of the high pressure, and passing it through the Steam Generator, which is basically just a large heat exchanger, on the secondary side of which is water which generates steam because it runs at a much lower pressure, and there you have it in a nutshell.

I did all the qualifying and eventually made Chief of the Watch, and then eventually Chief Tiff of the Boat, which gave me great satisfaction. Shortly after this last event we took the Boat round to Rosyth for an immediate Dry Docking, the reason we found out later was the fact that an American Skipjack class had a near-catastrophic failure of hull welding. As a consequence the engineering department spent endless hours ultrasonic testing with inboard boffins, and installing hundreds of strain gauges.

Eventually the modification and the general refit were over and I was all set to carry on with my normal duties after a manic and exhausting period when I had some bad news. A more senior Chief Tiff was appointed over me, which brassed me off more than a lot, even more so because on enquiring of the Chief Engineer why, he couldn't give me a satisfactory explanation, but insisted it was nothing to do with my performance. So I immediately slapped in for a draft because I didn't get on with the new bloke, but I was told that I had to give him a good handover period, which teed me off even more. This made up my mind not to sign on for another ten years, another reason being that home life was a bit rocky. So ended my Naval service!!

CIVVY STREET

Garage

When I first came out of the Royal Navy my Dad wanted to sponsor me in a Garage as cars was my hobby but I was a bit skint what with one thing and another. I discussed it with my ex, who was keen. I'm afraid I wasn't quite so keen as I knew that all my Dad wanted was a profit. But I got talked into it and eventually rented a Fina property with a Petrol, Oil and Accessories shop and workshop and hydraulic lift alongside and at the back of the shop. I employed a fitter and an apprentice, but the apprentice was better than the fitter who was more than a bit devious, but we did a good trade in the workshop and made good money. The problem was that the more money I made the more Fina put the rent up which was self-defeating.

Nevertheless we persevered and I learn a lot, but a few repair jobs stand out as challenges, but one in particular stood head and shoulders above the rest.

This particular customer came in and asked if I could help him. He explained that he had an Austin 1100, the one with Hydrolastic Suspension. He said it was running all right but it kept making this funny noise and it was beginning to get on his nerves. So I booked it in for the beginning of the next week. The owner only lived across the road from the Garage and it was duly delivered to us early Monday morning.

My fitter took it out for a test drive and came back shaking his head. He reckoned that there was definitely a very funny and very annoying noise but it was intermittent. He suggested that I came out with him to see what I thought about it. It was definitely an annoying noise almost like a very small bell tinkling away but only under acceleration or braking and it seemed to emanate from different directions.

We put it up on the lift and inspected it from stem to stern. It was thought that it may be an engine mount as, being a transverse engine a rubber mount may have split and may have been rubbing metal to metal but on closer inspection it was ruled out. The next suspect was the engine stabiliser, a small bar with rubber bushes to stop the engine rocking fore and aft under acceleration and braking, but that also was ruled out. The snag in this situation is that you can't charge money for something you haven't found.

I was underneath the thing again, tapping any anything and everything to try and replicate the noise. Then by accident I caught the exhaust pipe with the inspection lamp wire guard and it sounded a bit familiar. So we tapped away at the length of the exhaust but didn't discover anything conclusive. So I decided that we would take the full length exhaust down and give it a good looking at.

The exhaust was all one piece as original. It had the tail pipe and silencer at one end supported on rubbers and up front a right angle bend that terminates in a flared end with a two bolt flange which is difficult to seat squarely on the manifold and make it gas-tight. It is also clamped to the chassis at the bend.

Down it came and as we tipped it down to pull the bend out, there was the noise. We got it down on the deck and tipped it back and there was the sound again, we turned the right angle bend down and tipped up the silencer and the noise again and out popped a rather large lump of hardened Gun Gum, a paste that is applied to the flared end when it leaks at the manifold joint. Somebody had been more than generous with the Gun Gum and it had hardened under the exhaust heat and broken off and ended up zooming up and down the length of the exhaust.

How can you charge a lot of money for such a stupid fault.

I eventually jacked in the garage as I was by then paying more rent than the profit I made, much to my Father's annoyance, but even he had to eventually admit defeat.

Job Search

I'd left the Navy and failed at running a garage and had got myself a dead end job with a small engineering firm making drawings, in special ink, linking up small components for a magic-eye-controlled oxyacetylene cutter to follow. The boss was a prat who didn't trust anyone, and his Foreman was a snivelling wretch who'd shop anyone on the slightest pretext, but the job paid the mortgage and bought the food.

I was therefore scanning the local rag Situations Vacant section avidly, trying to get a better job, and eventually a couple got me interested so I applied for details on both of them. The replies I got back were encouraging especially the one near Selby who wanted a Chief Engineer on a mushroom farm looking after the sterilisation equipment of the growing medium among other things. I had an interview and was very optimistic of the outcome. The other job was described as Deputy Works Engineer at a chemical works processing animal and vegetable fats which didn't sound as interesting, but it was just down the road from where we lived. I also had an interview with them and it seemed to go very well with the MD calling in halfway through the interview who asked me about my Naval background and all about water chemistry which I think impressed him. The Chief Engineer then offered to show me round the site which was huge and filthy. The deck was covered in sticky material and as I wore slip-on shoes, was in fear of losing them.

The site generally was a shambles and messy, and the Boiler House was an utter disgrace and I was glad when the Cook's tour was over.

I received a polite reply from the mushroom farm telling me that I had unfortunately not been successful in securing the job, which seriously depressed me as my present job was getting more intolerable by the minute and I was getting desperate.

A couple of days later I received a letter from Croda regarding the position of Deputy Works Engineer, asking me to call again at my convenience, so I arranged an appointment. When I attended the appointment it was with the MD and the Chief Engineer who offered me the job with a very generous salary.

Although I didn't fancy the job I was desperate to get away from the job I had, and the money was brilliant so I accepted, and it turned out to be the second best job I ever had, the best one being the Navy.

As a postscript, not long after I started with Croda the *Hull Daily Mail* ran an article reporting the collapse of the Mushroom Growing Firm near Selby. Sod's Law, eh!

Gassed

I joined Croda as Deputy Works Engineer at the time of the threat of Edward Heath and the three day week. The MD at that time organised the purchase of a diesel generator and control panel. My gaffer, the Chief Engineer, decided that as a new project he would take charge of the installation himself.

He was an ex draftsman who had worked himself up and became an expert on fatty acids and their derivatives and he was very good, but knew nothing about diesels or generators. I was heavily involved in the steam generating side and rehashing the Boiler House generally.

One day a messenger came looking for me and I was told that the Chief Engineer would like me to join him in the Generator Room. In fact they'd cleaned out an old pipe store next to the Electrical Substation and Distribution Board which would be handy for connecting the Generator to the Distribution Board. I found the Chief Engineer peering in at the slip ring housing of the generator. He asked me to have a look at the slip rings so I looked in and shone a torch and recoiled in horror. The slip rings were steel and were deep solid rust! I asked him if he'd checked the windings for damp and he said he hadn't but he would. I told him that I thought it must have come into Hull as deck cargo after a long and wet voyage. The windings were checked and proved to be soaking wet, and I suggested electric tube heaters to 'slowly' dry things out.

In the mean time I suggested that a turning tool holder and feed carriage be rigged on the slip ring housing and the engine run up slow speed to skim the slip rings. He asked me if I would fancy performing the operation for him, and I told him it wasn't a problem as long as he didn't mind the Boiler House being delayed slightly, which he agreed to.

I got one of the fitters to help me flash up the Diesel, a V-12. It took a lot of turning over and venting individual injectors before it started firing on a few pots, half the trouble being a temporary fuel supply tank which was too low, so we lifted it up a few feet which helped. Eventually we had it self sustain running but a bit rough and found a few fuel leaks that we nipped up, but two injector pipes were split and had to be renewed.

Eventually we got the engine running fairly smoothly on Idle and managed to turn up the slip rings and made them smooth and polished them. But what I did notice was that by the time we'd finished the Fitter and I were feeling a bit rough, then I spotted why.

The exhaust had for some inexplicable reason been routed out through the front wall right above the radiator with the radiator fan drawing exhaust fumes back into the building. We were slowly being gassed and poisoned by carbon monoxide and that was just at idle speed. It would have been lethal at full chat. I went back to the Boiler House job having pointed out the error to the Chief and he had it re-routed through the roof, but whoever designed the exit shroud didn't leave sufficient clearance and once up and running at power the roof beams caught fire.

A few other disasters hit the installation but they are other stories.

Pig's Ear

Back with the diesel generator again with all its trials and tribulations. We had the unit up and running and on load which was keeping the Boiler House generating steam and half the mill running and producing in spite of the complete mains power shutdown.

The problem that developed was that the free end of the engine had an hydraulic vibration damper flywheel fitted which was driven directly off the crank shaft. It was therefore fitted with an oil seal which unfortunately started leaking and was beginning to make a mess. I organised a twenty four hour watch keeping rota. The watch keeper at the time, and the one who reported the leak to me was Dennis, an electrical contractor who had installed the control panel and wired it up to the switch board, and made a good job of it. I had a tray put down to try to contain the leakage but it wasn't working very well. So as the engine had quite a lot of ground clearance I asked the fitting shop Foreman to send me a couple of fitters to help me rig a better leak collection rig.

He sent me two ex-Gas Board fitters who were not noted for their geniality. I asked them to take dimensions and go over to the tank shop, the home of the welders, and get them to knock up a Pig's Ear! They looked at me a bit round-shouldered and wandered off looking more than a little confused.

I later learnt that they'd gone back to the Fitting Shop Foreman, and told him that I was taking the piss, because I wanted them to make a hash of the job. The Foreman sensed that something wasn't quite kosher and asked them to relate exactly what I had said and as soon as they said to make pig's ear he started laughing. He had been an HO in the RN in the war and explained to them that a Naval Pig's Ear was half a funnel, and not a cock up as in Civvy Street.

So away they went to the tank shop and gave the dimensions required to the welder and a pig's ear was made and fitted to the engine free end at the next shutdown and a pipe led the leakage to a container that could easily be changed for an empty one when the one in service was full, and no mess. The one thing the watch keepers had to do was to keep a careful eye on the sump oil level and top it up as required because of the leak.

But the two ex-Gas Board fitters tended after this incident to treat me very warily and tried to avoid me like the plague.

Oil Leak

The diesel generator was still running and producing electricity to keep the mill ticking over, but we were tending to lurch from one snag to another. One of the first problems we had was controlling the frequency as the speed adjuster was on the governor which was situate in the centre of the red hot engine. I discussed the problem with Dennis the electrical firm sub-contractor who was helping with the watchkeeping. I reckoned that I could rig up a very slow revving reversible modulator motor mechanically to the governor speed control if he could connect it electrically which he assured me was no problem, and before we knew it we were able to control the frequency from the control panel.

The next disaster was the lubricating oil cooler which was rigidly connected to the vibrating engine and started developing hairline cracks.

We used rubber insertion and jubilee clips to contain the leaks, but they could propagate and cause complete failure. I took dimensions and a had a firm I had used before to make up high pressure flexible hose connections and at the next shut down we fitted them, and not before time because as we disconnected the rigid pipes one of them just fell apart!

The diesel generator oil leak that we had tried to contain was getting worse and at the very next mains supply period it was going to have to be shut down and attended to, if not before, because it could really let go at any minute. It got to the stage where the pig's ear was not containing the leakage any longer.

The problem was that the oil leak was now impinging on the vibration damper flywheel and consequently the leak was now being sprayed all over and we had to rig canvas screens to try to contain it and the mess. I explained to the Chief Engineer that we had to shut the engine down to effect a proper repair. He decided that we would have to get the MD's permission so of we trooped to his super doper office, and Tom tried to explain to the gaffer what the problem was, but he didn't do very well and the gaffer decided that he had to see for himself, so he donned his immaculate white lab coat and off we trotted. He piled straight into the generator shack and tried to hold a conversation which was impossible because of the noise. I beckoned him outside and explained that the leak was being contained by the screens. So we went back in and I pointed to round the screen expecting him to just peer round the corner, but not him, he strode straight behind the screen and quickly reappeared absolutely covered in a spray of black lub oil, hair, glasses, face, lab coat, designer shirt, cavalry twill kegs, the lot. As he headed back to his office red faced and in a hurry he growled 'Shut it down!!'

Injured

The Boiler House was an utter mess when I was charged with putting it into good order.

A whole load of redundant pipe work, valves and general scrap had to be shifted out, and I installed proper boiler water sampling gear and I condemned the Boiler Water Treatment Plant as it was less than useless.

I had a new Dealk Degass Base Exchange system installed which supplied good quality treated feed water to the boilers. Unfortunately the legacy of neglect persisted as a lot of limescale had been laid down previously.

In a shutdown period where time was limited I would organise a high pressure jetting crew to blast as many boilers as possible, but the big high pressure Babcock and Wilcox boilers were a real nightmare because every now and again a big four-inch bore generator tube would rupture and they always split on the underside where the flame impinged, and which is where the worst limescale builds up.

When a tube ruptures the four hundred pound per square inch pressure reaction force bends the ruptured tube up till it hit the next row up and it is restrained. The boiler repairers were then called in to replace the damage tube. A stock of tubes was kept on site and the repairers were fast and efficient as they'd had plenty of practice.

The situation regarding tube failures was reducing due to the efficiency of the new water treatment plant and most of the redundant pipe work and scrap had been removed, so I decided that it would be a good idea to decorate, and tart the place up a bit, just a decent coat of nice cream emulsion, and colour coded pipe work to readily identify same. I recruited the Boiler Men and Water Men to do a bit at a time while they were on shift in between attending to their normal duties and I assisted to show willing. It took a bit of persuasion but they all eventually joined in, even the usually bolshie Union Shop Steward. Finally the decorating was completed and although I say it myself it looked a picture compared to what it had been and a lot of blokes big and small congratulated the lads on their efforts.

At the time of the accident Fred was the duty boiler man and he was on the firing floor taking his hourly readings when there was, he said, an almighty bang and suddenly the heavens were raining scalding hot brown water with bits in it and bricks!

Fred was injured, he hit his head on the cross beam above the ladder he was escaping down to the lower level away from the falling junk. Of course the boiler shut itself down but the damage was done and it was horrendous, the reason being that it was a top row tube that had blown, the first ever.

The reason for all the damage was the very fact that it was a top row, and instead of being restrained by the next row up, it had flipped out of the sinusoidal header due to the reactive force and took out the top steel support place and all the brickwork on top of it, and some of the side wall brick work as well. It even blew a few of the roof sheets to kingdom come.

The annoying thing from my point of view was that all our hard decorating work was ruined. Most of the surrounding areas looked as if it had been pebble-dashed with dark brown boiler water and limescale pieces.

All Fred needed was a sticking plaster, having avoided all the brick work and junk raining down, and I had a Portakabin installed on the firing floor between the Babcocks as an office and refuge in case of any repetition.

Babcocks

The Boiler House in total generated 65,000 pounds per hour of steam of varying pressures the most used being for 100 psi generated by two Babcock and Wilcox water tube boilers. They were originally coal-fired then, with the advent of natural gas which was not only cheaper but cleaner and less labour intensive, the chain grate and all the coal feed gear was ripped out and converted to upshot gas burners, two per boiler. The secondary fuel was gas oil (diesel) but they didn't work very well and tended to fractionate the fuel and the residue used to collect in the burner base and eventually flood and extinguish the flame.

Someone very high up in the firm decided that we should burn our own residues generated from the processes. We at the sharp end had no say in the design or construction or even the materials of construction, which caused all sorts of problems in the future.

The design was that the fuel lance was fired horizontally through the front fire wall and a ramp of fire bricks built in the first firing chamber to divert the flame vertically through the generator tubes.

A holding and mixing tank was built in the ex-coal yard and the residues were collected and mixed in accordance with the laboratories recommendation, and was kept liquid by a steam heating coil. When required the oil residue is pumped to a steam heated heat exchanger in the Boiler House base which of course increased its temperature and reduced its viscosity to enable it to be increased in pressure by three cylinder positive displacement pumps and fed at high temperature and pressure to the multi-orifice cap on the end of the lance and atomised by high pressure steam and fired up initially electrically, and was then self-sustaining. We were then generating high pressure steam from our own rubbish!

The problem was that nobody told the designers and manufacturers that the residues their equipment and handling was very acidic and very abrasive. The materials of construction of the wetted parts was carbon steel which was fatal as the fatty acids just ate it. It even slowly ate 316L (low carbon) stainless steel. The first casualties were the Lance atomiser caps which believe it or not lasted on average ten to twelve hours before the holes themselves had enlarged and tended to link together. I've even see one where the whole of the centre had completely disappeared.

I recruited by best Turner to replicate the cap design in 316L stainless steel which after a couple of false starts he perfected and lasted infinitely longer than the carbon steel jobs. The next bits to fail were the valves of the HP piston pumps and stainless steel had to replace the originals. By this time the gaffer of the installers, a laconic Scotsman, was beginning to have his doubts about the whole installation.

The next item to fail was the steam heat exchanger where the tubes started rupturing because they were slowly being eaten. A halt was called to the whole project and the Contractors went back to the drawing board and redesigned everything that came into contact with the residues, and were all produced in best quality stainless steel or hard faced materials which although was still quite labour intensive lasted a lot longer than the carbon steel alternative.

The one snag that nobody expected was that after a while the combustion chamber started to choke up and the boiler lost efficiency and had to be shut down and it took absolutely ages to cool down sufficiently to be able to survey the problem. We were all astounded with what we found. The floor of the chamber was full of what can only be described as black glass and when broken up with pneumatic chisels was as hard and sharp as glass. A sample was sent to the lab for analysis and was found to be rich in Nickel, which was used as a catalyst in one of the refining processes. A firm was found who would process the solid residue to recover the Nickel which was an added financial bonus!

As the old Yorkshire saying goes, 'where there's muck there's brass'.

Spam

With the help of the boiler men and water tenders we managed to get the Boiler House into reasonable order. The feed water quality was good and we were also taking every opportunity to descale any boiler that became available.

The feed water header tanks had been cleared of sludge and the internal deck re-plated, and loads of redundant pipework removed. The big things that bugged me were the remains of the coal-burning era. The Telfer gear and the overhead crane system which transported coal from the external coal yard to the internal feed hoppers. All this just slowly rotting, and the coal yard filled with pallets loaded with redundant and scrap forty five gallon drums most with no lids so that rain water has filled them and flooded the fat contents out onto the already messy area, a proper shambles.

All of a sudden there was hell on throughout the mill with Police crawling all over the twenty- acre site. It appears that a storage warehouse next door to our site had been broken into and some food items had been stolen and it appears to have been going on for some time. Some damming evidence such as a broken and disguised perimeter fence on our boundary pointed the finger firmly at out site.

My usual routine first thing in the morning was to have a word with the boiler man to check that everything was OK. I climbed the steps to the firing floor and noticing a sticky substance on the tiled deck. On further investigation the substance was found to be some sort of paint, red lead or similar and it appeared to be dripping from the discharge valve of one of the coal hoppers, so I climbed up the dusty ladder and forced the top door open on rusted hinges and came out onto the walkway which led to the parked Telfer Carriage.

Looking into the dripping hopper that had assorted rubbish in the conical bottom was what looked like an overturned paint tin with something up against it.

I went below and donned overalls and, armed with a length of rope, an extending ladder and a couple of day fitters. We trooped back aloft, rigged the ladder and with the assistance of the rope, descended to the rubbish at the bottom of the hopper.

The tin indeed contained red lead paint which had been tipped over by of all things a tin of Spam, not just a little oblong supermarket variety, but a big long round catering-size tin.

The outcome of all this was that more investigations around the site uncovered all sorts of tinned goods, abandoned all over but the majority was located in rain filled redundant barrels in and around the coal yard.

The Police investigation revealed that the main raiding party on the Storage Depot was from the Boiler House night shift and adjacent plants. A few sackings and prosecutions ensued, and it took me some time to train up new boiler and water men but the good thing from my point of view was that I recruited and trained them to my standards and satisfaction, and the steam raising facility was in my opinion more efficient and reliable.

The Column

To process fatty acids you have to remove the glycerols from the feed stock, be it tallow, rape oil, and sometimes fish oil, the process being called Splitting. The old, still used, method was with three high pressure vessels which processed the feed stock in batches which was not only costly, but also slow and inefficient.

By this time I'd been promoted sideways to Maintenance Engineer which suited me down to the ground as I had an office down the Fitting Shop and I'd got rid of all the big-wigs and Plant Managers and I could do my own thing.

The top management had realised that the splitting capacity was a bottle neck, and had ordered a hundred foot tall continuous process splitting column from Lurgi in Frankfurt. Unfortunately our site was ex-marshland so lots of piles had to be driven to support the weight of the column as it worked at very high pressure and consequently was very heavy. After the piles were driven a reinforced concrete base was cast around the pile caps and studs embedded in the base to the number and the dimensions of the column base flange.

The process was thus, feed stock at high temperature and pressure was pumped in to the bottom of the column and hot high pressure water was pumped in at the top. The feed stock which is fat, floats on water, so the feed stock changes places with the water. The water on its way down collects the glycerols from the feed stock and is discharged at the column bottom as sweet water and is evaporated and filtered to produce glycerine. The fat discharged out of the column top is crude fatty acid and goes on to be further processed and refined.

We were eventually informed that the column had been dispatched and was on its way via North Sea Ferries. The entrance to the site was fairly narrow and articulated tankers found it a tight corner so the entrance gates had to be removed and also the near gate post and the surface at that corner laid with steel plates.

It was planned to move the column from the docks to site early Sunday morning while traffic was light, and we had two cranes on site, the main one being built from four lorry loads of bits. The other one was a very large telescopic unit to be used as the tailing in unit, which entails manoeuvring the column vase for the main crane to take the full vertical weight.

To cut a long story short the column eventually arrived on site after a couple of hiccups. While it was parked horizontally on site the Lagers took advantage and did a good job on it.

The big crane was by now fully assembled and early doors the big day arrived when we were going to lift the beast into place. The snag was that the weather forecast was not good, but any delay would be very costly with crane hire and I was told that the lift would go ahead no matter what.

So be it, I had the Foreman Fitter help me direct the cranes and eventually grounded the column over the studs and the fitters started bolting it down but the wind was getting up and it started to rain and came in very dark. I took John, a newly qualified fitter, with me and we climbed the four stage ladder to the top of the column to unshackle the crane. We were half way up when the heavens really opened and it started with thunder and lightning. We eventually gained the column top walkway and started in release the lifting tackle with the thing swaying badly in the wind, almost as if the securing nuts were still loose. Then suddenly there was a flash and a bang and we both saw a lightning strike in the sports field next door to us. I was thinking a little prayer when John piped up and asked me if we were safe and I reassured him that everything was A-OK. Knowing full well that the lightening conductor was not yet connected, and we came down to ground level fast. We were freezing cold and soaking wet through but glad to be safe.

After a few false starts and a slight internal redesign the continuous splitting column performed absolutely magnificently and very efficiently.

New Pump

The North Pump House was one of the places that kept the mill running by supplying cooling water mainly for maintaining all the vacuum plants on site. But the Pump House itself was a shambles. All the pumps were obsolete, most of them LE3 split casing low revving large diameter cast iron types which were very high maintenance because spares were not available. Most of the shaft glands were worn out and were peeing out leaving the desk awash with river water and mud, not a very nice place to work in.

The system worked thus, two of the three abstract pumps ran twenty-four-seven, the third one being standby, drawing muddy water from the River Hill and discharging it via a float level control valve and strainer basket to two very large concrete tanks. Large diameter pipes supply this water to the suction side of the North Pump House. Each plant that required cooling water had a pair of supply pumps, one running, the other stand by in case of failure, which was far too frequent and as I've already said, all the pumps were in very poor condition and kept fitters and turners busy rebuilding them which was very time and money consuming.

I came across a pump supplier who had just come back from the Middle East after making a killing selling Ryland pumps to the oil industry. He had just set up a business working from an old ex-church in Beverly and had made an appointment to come on site and give me his sales pitch. He sounded as if he knew his pumps so I took him to the North Pump House and we paddled in to show him the muddy water shower because as I've said, all the running pumps had badly leaking glands.

He was not impressed and he showed me the blurb on a range of cast steel pumps, high speed, small diameter volute casing, with flexible drive and pull back impeller which sounded ideal to replace our obsolete heaps of junk.

I requested and got permission to spend money on one unit to test the types suitability. A mounting bed had to be fabricated by one of our welders in the Tank Shop, and the suction and discharge connections modified to suit with new inlet and outlet valves. Come the big trial and Tom the Chief Engineer and Dave the supplier were in attendance to witness the trial.

This trial pump had been fitted to supply the biggest Fractionate plant on site so I had my fingers firmly crossed. The pumps were changed over and the discharge line gauge showed a healthy pressure and it was running so smoothly and quietly and we were all about to congratulate ourselves when suddenly the pressure dropped right off for no apparent reason, and the pump we'd just changed over from had to be restarted. Poor old Dave looked devastated and I didn't feel too good about it either. I asked the fitters who were standing by to disconnect the coupling and draw the impeller back to inspect it for any faults. This they did and lo and behold the eye of the impeller was completely blanked off by a very large eel that was wrapped neatly in impeller centre. As soon as the fitted pulled the offending eel out it came alive and so was returned to the river to fight another day. The pump was reassembled and ran a treat for years and slowly, as the maintenance budget permitted all the old pumps were replaced, much to everybody's satisfaction and the fitters when they were called to the North Pump House didn't have to paddle anymore. And Dave secured himself a nice little earner. To prevent a repetition of the eel incident a wire mesh was fitted over the filter basket to stop them leaping out into the concrete tanks and the inevitable consequences.

Collapse

One of the biggest plants on site was Frac 3 a two column Fractionate Plant.

The feed stock was crude fatty acid and produced highly refined almost one hundred percent pure cuts of various carbon molecule chain lengths.

This plant was designed and fabricated by Lurgi of Frankfurt and shipped to Hill and erected and assembled by Croda Fitters, Welders and Electricians on direction from Lurgi Erection Engineers.

I joined the company after the plant had been running for about five years. Lurgi predicted a lifespan of ten years maximum. In fact it was kept running for more than three times that amount, but not without problems, most of these being with the first pass larger column. Both columns had multiple bubble trays where the more volatile vapours bubbled through the liquid in the trays. The liquid eventually arrived at the bottom of the column and was pumped back to the top via a heat exchanger to be reheated by high temperature thermex that had been heated to a precise value by an external gas fired burner. The fraction required was drawn off the top of the first column via a large vapour line for the process to continue in the second, smaller column. But it was the first column that was causing concern as its output was reducing for no apparent reason.

Eventually it fell off so badly that the whole plant was shut down and cooled down. When sufficiently cool the vacuum was broken and the only access to the interior, a manhole on the working floor, was opened up and a fitter and myself entered, after doing the mandatory atmosphere checks, to do a survey. Each tray had a closed access panel and the higher we got the worse the conditions of the trays got due to corrosion.

Then about half way up we couldn't go any further as the trays had completely collapsed on one another and all the trays and caps were as thin as tissue paper, and as we evacuated we had to be careful not to pull any more trays down on top of us.

A good fabricating firm was called in and employed refabricating the trays and caps from Lurgi's original drawings, while we removed all the collapsed and dodgy-looking internals.

Eventually the plant was restarted and production was back to normal.

But one thing worried me because at start up the lagging on one side at working floor level started smoking. The lagging was over a foot thick and sheathed with sheet aluminium and there was no other reason that I could see for the smoke other than a leak of fat from the column into the lagging while the vacuum had been broken and still warm for the fat to be liquid.

I kept my fears to myself but kept a careful eye on things, but I wasn't too worried as the vacuum was still being well maintained so the leak couldn't be too bad, but one thing I noticed as I took the lagging cover off the manhole and looked through the inspection port hole, I could see the material discharged from the heat exchanger shooting out horizontally and impinging on the far side of the column right where I'd seen the smoke.

I kept my trap shut because the summer shutdown was only a week away and the whole mill shuts down for a fortnight, which is when my maintenance department go to town on all the major defects and insurance inspections.

But my main concerns once shut down and cooled was Frac three first column and got the Lagger to strip the area where the smoke had emanated from, and to our complete amazement the area that had been uncovered had been sucked in, the shell was not only pulled in because it was so thin it was also perforated with lots of pin holes, which is where the leak and therefore smoke came from. I showed our findings to the gaffer and he was horrified and as I explained to him my main concern was the fact that the whole of the columns weight was supported on the working floor which was below the weakened section and I pointed out that the whole column could collapse at any moment unless we were very careful. He went away wringing his hands and shaking his head.

First off we had to be sure that the column was still vertical so a section of lagging at the top was removed and the shell checked with a very sensitive spirit level and was found to be still vertical thank goodness. The next step was to make sure it stayed vertical so I had an RSC welded to the good bits top and bottom in the centre of the bulge. The next stage was to renew the whole of the base section shell and the only way to do that without the whole thing collapsing on us was to replate piecemeal.

A narrow strip at the thinnest area was removed using a plasma cutter, the reason being that this type of cutter didn't produce much residual heat and therefore little distortion. A new section the size of the piece cut out was cut and rolled to the right diameter and welded in top and bottom. A similar sized section was cut out next to the new piece and another new section was welded top and bottom and then welded vertically in the first new section. The other side of the new first section was treated similarly and continued all the way round until the whole lower section was fully restored without the whole damned thing collapsing on us which had been a real concern.

As a preventive measure I had a thick disc of stainless steel installed as a sacrificial target to stop the far side of the vessel being eroded so badly.

We now know the general condition of the plant and in the future a smoking area on start up spelt leak and operators became more vigilant and smoke became more frequent. The fear was that it might catch fire and we had a couple of small conflagrations in my time that we managed to contain, but shortly after I retired I heard that the whole thing had burnt to the ground which in my opinion was inevitable, but thankfully no one was hurt.